Anatomy of a Park

Fourth Edition

Anatomy of a Park

Essentials of Recreation Area Planning and Design

Fourth Edition

Donald J. Molnar, FASLA
Purdue University

Long Grove, Illinois

For information about this book, contact:
Waveland Press, Inc.
4180 IL Route 83, Suite 101
Long Grove, IL 60047-9580
(847) 634-0081
info@waveland.com
www.waveland.com

10-digit ISBN 1-4786-2202-4
13-digit ISBN 978-1-4786-2202-4

Printed in the United States of America

7 6 5 4 3 2 1

ET AL

For Albert J. Rutledge, author of the first edition

Anatomy of a Park is a gift and a guide, not only for "parks persons"
but also for folks who are learning about land and its use by and for people.
Al watched people because he enjoyed seeing them having fun—*being people*.
His scholarly and witty knack for verbal description made the original text come to life.
I drew sketches for the new editions with him in mind, because I know how he enjoyed drawings.
I enjoyed his company. I miss him. We all do. Thanks, Al.

Contents

Introduction

This book lays bare the essentials of park design. It is written primarily for three major groups:

- *Non-designers*—park directors, park board members, recreation leaders, park planners and managers, *and anybody affected by the designer's product—the park.*
- *Park designers*—landscape architects, architects, engineers, anybody who tackles the task of creating a park from idea to execution.
- *Faculty and students*—the people who are teaching or learning to be future practitioners of any of the foregoing professions.

Anatomy of a Park has a long and happy history of shared wisdom. The Fourth Edition maintains the basic structure and adds much in contemporary information from direct interviews with professionals in management, design, and construction. To better appreciate the scope of the book, it's helpful to see it as a series of zones to learn the language of *Park*.

Zone One—How to Think *Park*

Chapters 1, 2, and 3 are still repositories of values, principles, and aesthetic understandings. The *umbrella considerations* have received a larger umbrella covering issues of green design (beyond leaves and grass) and sustainable thinking.

Chapter 4 maintains the *functional-considerations* focus on relationships of uses and the reality check of how things work. Recognizing the importance of seating in any park, introduced in this chapter is the *Encyclopedia of Seating*, which serves as a mini-text (with maxi-illustrations) of the right ways and wrong ways of arranging and designing the site furniture which gets more use than any other equipment in the park. Further exploration into site planning includes consideration of the rapid development in the linking of paving and storm water management as well as issues of survival for plants in dense urban environments. The essential element in Chapter 4 is a focus on *getting more park for your money*.

Zone Two—How to Read *Park*

Chapter 5, Plan Interpretation, introduces examples of the graphic language used by LAs to explain the three-dimensional reality of a park, from the existing description of the land as a topographic map, to an aerial view or a ground-level perspective and all the types of plans brought into play as a project develops. New in this edition are truly 3D views of the topography, as though built in model form, to help viewers comprehend the meaning of the squiggly lines with numbers that are used to describe the whole world.

Zone Three—Applying What You've Learned: Process and Appraisal in *Park*

The most challenging and most important zone in *Anatomy of a Park*, chapters 6 and 7, provides step-by-step introductions to the process of site planning and the analysis and evaluation of park plans at all scales. New in these chapters are three-dimensional models of some of the sites, bringing the topography to life in some and showing the spatial essence of others.

Chapter 6, Site Design Process, shows the steps in preparing to design: the inventory of information available; the analysis of the information; the synthesis, pulling the information together to listen, graphically, to its suggestions; the appraisal of relationship diagrams; design concepts; and finally a refined plan. Although there are decisions being made, guiding toward a commitment to a final result, you're seeing how to make those decisions . . . without having to make them—*yet*.

Chapter 7, Plan Evaluation, is the off-the-deep-end experience in which the basic process is gathered in a detailed checklist you will use to evaluate a series of case studies at scales ranging from a tiny parklet to a huge regional park. The importance of using this checklist is that you will gain the ability to see backwards in time from a finished plan to evaluate the preliminary development of solid support for the conclusions represented in the final plan. With the preparation of Zones 1 and 2, Zone 3 represents the inoculation to prevent a response to the presentation of a fancy plan from being either a shallow "Ooh, I like it" or an unconsidered "Naah, I don't like it." This chapter is a challenging but very rewarding experience. It's a great teaching tool, a disciplined refresher, and a very useful basis for a dialogue (or argument) between people learning to make decisions about a very expensive and critically important investment: a *park*.

Zone Four—Making It All Happen

Chapter 8, on hiring the right designer and being a great client, is a new chapter intended for the client seeking a consulting designer as well as the designer seeking a client. Based on information gathered across this country from professionals in both roles—clients and consultants—their good and not-so-good experiences help distill a package of guidelines and procedures that cover everything from the earliest initiation of a park project to the ribbon cutting and life following.

Essentially, the focus is on client *with* consultant, not client *against* consultant. Seen from viewpoints of both parties, the process of selecting the right designer is a measure of a successful park system administration. Being selected, and selected over again, is a measure of a good design office. The information is valuable to both parties, whether at a first-time stage or long into careers.

Chapter 9, Good Park, Great Park, looks at what constitutes a great park in the minds of the ultimate critics: the users of that park. Where are they and why are they? And beyond what already exists, how does a community continue to grow a park *system*?

While Chapter 4 is aimed at *getting you more park for your money*, Chapter 10, The Bottom Line, is about *getting more money for your park*. It's all about what you have for sale (and it's not cookies), seeking money from government sources, gaining services from public support, and transforming public enthusiasm into personal effort toward building and keeping great parks. This edition includes ideas and suggestions for building partnerships that help communities conceive of and implement new parks and open space. My thanks to Bernie Dahl, Purdue University, for the original preparation of this chapter for the Third Edition.

Anatomy of a Park is and always has been a volume that views parks in totality, from the germination of an idea into its expression as a creation of experiences, with a life as long as the generations who enjoy it. People sitting around tables do not build parks. It takes a competent designer to assess the needs of the client, translate those needs into a comprehensive plan, and communicate the resulting ideas to park constructors and managers. Likewise, parks don't just happen. People in the community express their needs, park boards prioritize that community's needs, and city administrations provide the funds to build and maintain the parks.

Although there are several books out there that are either totally or partially devoted to park design, most of them also focus on such special aspects of development as playground design, marina planning, or forest recreation construction. Many of these volumes, along with the few that attempt to be comprehensive, are noted in the bibliography. However, despite all the writing that has been done on the subject, one element is missing: a basic manual on park design that ties the fragments together and helps make other readings more meaningful. *Anatomy of a Park* provides such a framework, to which the details of more specific readings can be related. It is written as a *nontechnical technical* book to assist the non-designer yet provide a substantial introduction to the field. *Anatomy of a Park* is a basic reference and teaching text on the fundamentals of park design. Augmented by personal experiences, slides, plans, and field trips, this book will serve as a guide for presentations of the essentials of park design to majors in the fields of park management and recreation programming. It can also provide organizational material for the preparation of short-course seminars for professional administrators and other agency personnel.

As with the first three editions, we have tried to provide information without a "use before" limit on the subjects that concern designers who plan lands for recreational use. We hope it will continue to be useful for years to come. For a basic briefing on park design and development, this is still a good place to begin.

Special credit is given to this book's first author, the late Albert J. Rutledge, FASLA, for his timeless wisdom.

Donald J. Molnar, FASLA

1 Context
Past and Present

We live in an age of specialists. This is good, for it allows subject matter to be probed and applied in depth. But specialization has its drawbacks as well, since many disciplines have become so specialized that their implications frequently cannot be understood by people in related fields. As a result, experts are often isolated from the potential contributions of others.

In today's complex society, no single discipline can pretend to solve substantial problems without collaborative assistance from allied fields. Accordingly, this book proposes to foster an understanding of design and designers among park administrators, recreation leaders, and others responsible for park development. All of us must perform as a team if quality works are to become widespread.

While it may seem reasonable that a team relationship should exist, it is missing in too many cases. To some extent, lack of collaboration is due to the professional isolationism brought about by the specialization phenomenon. Therefore, the history of the field warrants review and evaluation with an eye to its contemporary relevance.

From Olmsted to Today

It should be no surprise that park and recreation policy makers and landscape architects are often allied when one realizes that we have the same parent: Frederick Law Olmsted (1822–1903). It was Olmsted who, in describing his role as designer-superintendent of New York City's Central Park, conceived the title *landscape architect* and applied it for the first time in the mid-nineteenth century to those who organized land and objects upon it for human use and enjoyment. It was also Olmsted, with the development of Central Park, who initiated the first real park and recreation movement in the United States. (See Figure 1-1.)

Out of these beginnings emerged three related phenomena: an American design style suited to nature-oriented parks; a split of recreation authorities into two opposing camps—passive recreation enthusiasts and active recreation advocates; and a situation in which landscape architecture schools provided many park authority leaders.

In part, these traditions were rooted in a revolution that occurred in England preceding the development of Central Park. Nineteenth-century industrialized England was laced with dirty, vice-ridden, poverty-stricken, run-down cities. As a psychological counterthrust, England moved into its Romantic period, its people escaping from their oppressive environment through songs and poems that expressed nostalgia for idyllic nature and had a strong component of fantasy.

The well-to-do, utilizing their financial advantage, escaped physically, fleeing with their families and belongings to the countryside. There they called upon designers to plan their country estates. Sensing a desire for relief from every reminder of the city, designers strove to create patterns which excluded the axes, circles, squares, and other geometrical patterns which visibly organized the city.

They discovered the alternative in nature and began to lay out roads, walks, and other use areas in the "loose" organizational sys-

Figure 1-1
Frederick Law Olmsted (right), the nation's first landscape architect, designed Central Park (left) and supervised the early stages of its construction.

tems associated with nature, rather than with the regimented forms produced by the mathematically oriented minds of their predecessors. Sweeping lawns and meadows appeared. Human works were fitted to existing land configurations. Plants were allowed to exhibit their natural forms. As shown in Figure 1-2, the acreage took on a relatively undisturbed "natural" appearance despite the inclusion of constructed necessities.

Figure 1-2
The English Romantic style of landscape design.

Olmsted found himself in a situation somewhat parallel to that in England. In the 1850s, New York City was also industrialized and highly overcrowded. While Olmsted was sensitive to the "English solution," he was equally concerned with the plight of the common person. Reasoning that the entire population could not flee to the countryside, he proposed that there should be rural landscape within the heart of the city, where a person could go quickly to "put the city behind him and out of his sight and go where he will be under the undisturbed influence of pleasing natural scenery."[1] Working with this principle in mind, he collaborated with English architect Calvert Vaux in a competition to design Central Park. The Olmsted-Vaux plan won, and Olmsted stayed on to oversee the construction.

Olmsted was an extraordinary man, Central Park (the first planned park in this country) being but the first of his contributions to open-space planning and development in the United States. Subsequently Yosemite Valley in California was set aside as the first state park. Its preservation was brought about as a result of a report Olmsted compiled in 1865 on the natural wealth of the region. Shortly thereafter, Olmsted conceived the idea that municipalities should link a series of parks into a working complex, thereby evolving the concept of a park system. His influence in this regard is still evident in such cities as New York, Buffalo, Philadelphia, Boston, and Washington, D.C., the lands proposed by him for acquisition being the backbone of their park systems today. He was also the creator of the parkway concept, having advanced the idea of this type of roadway to connect Riverside (which he designed as the country's first residential subdivision) with Chicago in 1869.

Olmsted's superior ability to blend roads, buildings, walks, and other construction into the existing landscape without visible disturbance led to emulation by other outdoor-area designers, now called landscape architects. Although they provided a legacy we enjoy today, many became trapped by the style that had proved so successful for Central Park. While such naturalized developments were well suited to passive and semi-active types of recreation, they were decidedly more difficult to justify for facilities for active sports, especially when these were located in little corners of the urban scene where segments of unassuming nature might appear strangely out of place, if it was possible to install them at all. It was also argued that active sports and related activities hardly needed to be surrounded by rustic trappings. Accordingly, the landscape architects of that time and many from the twentieth century either shunned the design of active sport facilities or attempted to stuff them into a rural mold.

On the recreation side of the ledger, Olmsted was a persuasive champion of his cause. He was convinced that the essential recreational need was for rural retreats in the heart of the city. This conviction became the primary tenet of the park and recreation movement of the times. As the movement grew in acceptance, another came

[1] Charles E. Doell and Gerald B. Fitzgerald, *Brief History of Parks and Recreation in the United States*, The Athletic Institute, Chicago, 1954, p. 33.

along. Physical education enthusiasts proclaimed that recreational pleasure could also come from planned exercise, skill development, and the excitement of competition. Rather than cooperate, Olmstedian naturalists and active recreationists became wrapped up in the virtues of their own causes and formed poles not only of thought but of administrative authority as well.

Parks were defined as naturalized passive retreats. *Park departments* became solely concerned with parks as just defined. Landscape architects (schooled in the Olmstedian tradition) turned their energies to park development and, in their enthusiasm, included options in park management in their university programs. Such background led landscape architects in the early part of this century to roles as park administrators and policy makers, in addition to their role as designers.

Concurrently, *recreation areas* were defined as active-sport-oriented facilities—they included playgrounds, hard-surface court areas, and team sport fields. Municipal *recreation departments*, separate from park departments, were formed to handle only recreation areas and were staffed on the policy level by administrators with a physical education background. As a result, *parks* received much design attention but *recreation areas* very little. Unfortunately, one can only conclude that such separation and lack of interdisciplinary collaboration resulted from nothing more significant than professional pique.

This situation remained relatively stable until after World War II, when new demands pushed aside the old conflict. Population upsurge and unparalleled economic expansion focused eyes upon the question of land use in its broadest sense. Accordingly, landscape architecture education expanded its horizon in order to cope with the contemporary complexities of any type of land use problem: subdivision, campus, industrial complex, active and passive park, plans for whole cities and regions. To make room for new design programs, many schools dropped park management as an option for their landscape architecture students.

At the same time, population pressures and increases in leisure time raised serious questions about the depth and scope of the recreational experience. The same population upsurge hastened purchases of public parkland to the point that heretofore sleepy park agencies awoke abruptly to find that they had become complex managerial concerns. In reaction, university programs in park administration and recreation research were expanded, and many more were formalized as recognized entities with highly sophisticated missions.

Today we are left with vestiges of the historic hangover; some park departments are still not speaking to recreation departments. But these situations have become increasingly rare, for one happy result of the post–World War II trends was the unstoppable movement toward combining all leisure-time services under one administrative roof. This process was finally to force recognition of what should have been self-evident from the onset. Leisure-time needs can be satisfied through both active and passive means, as was implied midstage in the movement by Charles Doell in a few simple definitions. Doell, superintendent of parks emeritus of the Minneapolis park system, set forth recreation simply as "refreshment of the mind or body through some means which is in itself pleasureful"[2] and defined park in its broadest sense as a "piece of land or water set aside for recreation of the people."[3]

[2] Charles E. Doell, *Elements of Park and Recreation Administration*, Burgess Publishing Company, Minneapolis, 1963, p. 3.
[3] Ibid., p. 5.

Agencies have also witnessed an influx of university-trained administrators who have inherited the roles formerly filled by landscape architects, engineers, horticulturists, and others who once moved from their related interests to management posts. The intensive background in management possessed by new administrators has been gained at the expense of technical prowess in some areas, thereby increasing their reliance on on-the-job advice from others in such fields as design, construction, and maintenance. While much technical problem solving must now go the collaborative route, final approval remains as before in the hands of the administrators by virtue of their continuing positions at the top of the organization chart, along with the ultimate responsibility for the acts of their consultants and staff. How administrators and consultants work together successfully and productively in the highly specialized world of the present is probed in depth in Chapter 10.

Meanwhile, in recognition of the contemporary philosophy, landscape architects are paying equal attention to facilities for passive and active recreation pursuits. In the past few decades, mainstream designers have acquired the habit of designing land-use units on their own merits, having purged from their thinking the preconceptions that previously led many to stuff design elements into nonconforming molds. This sensitivity to the need for responsible land use was not the result of an accidental discovery.

During the 1960s and 1970s, the public visibility of both recreators and landscape architects increased. Generous federal funding became available for land acquisition and development, and agencies multiplied their holdings at an unprecedented rate. Unparalleled demands for increased quality of service came from a vocal and politically alert citizenry sensitive to the use of federal funds (their tax dollars) and anxious to participate in decisions about spending those funds. Environmentalists and behavioral scientists especially came to influence decisions about the provision of places that could have a salubrious effect upon the human condition.

Environmentalists, for instance, focused on the ugliness of both the city (Figure 1-3 on p. 6) and the countryside wrought by building booms with back-to-back developments characterized by blank reaches of asphalt, glaring signs, unpalatable forms, and choked roadways tangled around walking routes—all like dirty dishes in the sink (Figure 1-4 on p. 6). This raised a demand for visual refreshment as both a basic need and a fundamental human right. Behaviorists reinforced this notion with the supposition that physical surroundings consciously or subconsciously shape human attitudes, breeding tranquility or tension, pleasure or dissatisfaction (Figure 1-5 on p. 7). They surmised that the chaotic physical environment was adding tension to tension and thus exacerbating the already unbalancing stresses of job, home, and everyday existence.

It was noted that Olmsted's parks were for escape. But was escape entirely possible in an age when unlimited mobility took us past so much shoddiness on a daily basis? How lasting was the momentary relief of a park experience if one had to return to face that which drove one to escape in the first place—when what drove one to escape comprised by far the greatest proportion of one's environment? The prognosis was discouraging. Accordingly, parks were given a larger purpose than vehicles for escape. Rather, it was postulated, they should be developed as exemplars of what is possible in terms of soul-satisfying environment both for those who engage in an activity within and for those who pass by daily. Parks were also to be cata-

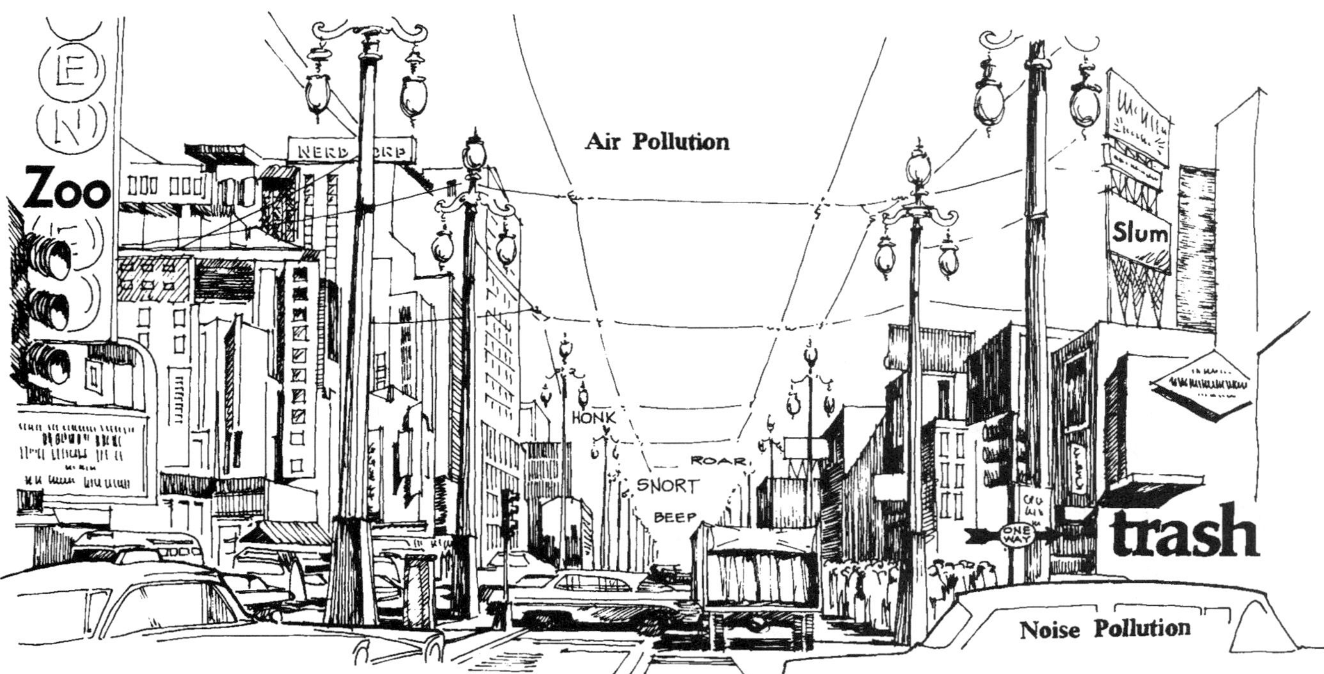

Figure 1-3
Blight in the city.

Figure 1-4
Chaos in the countryside.

lysts for promoting higher standards in other types of development, helping to move us toward the day when everything human beings build will contribute to positive physical surroundings.

Mental exercise was, it was also argued, another need of the times. Mass production systems that provided a wealth of goods had at the same time produced an environment of repetition. Uninterrupted monotony dulls the brain and creates mental lethargy. On the other hand, mental capacities may increase when the mind is continuously called upon to act. When brains are subjected to difference, comparisons are made, capacities to distinguish are exercised, curiosities are developed. To counter the phenomenon of creeping boredom, designers were called upon to provide environmental diversity by ensuring that each development was distinctive and unique.

Mass-produced products were additionally seen as contributing to identity loss, for they were tailored to the "average" person. However, "average" can be a misleading concept (as in the case of the person with head in the oven and feet in the refrigerator who concludes that "on the average, I feel comfortable"). The average person does not exist. Thus, products created for that person are created for no one in particular.

Pride, pushing aside the hollow feeling of anonymity, can result from being surrounded with things that have personal associations. This notion gave urgency to the need for activity-preference surveys and other forms of citizen expression to ensure that parks were developed according to the actual needs of user populations rather than those of mythical average people. It also mandated designers to attempt to give developments "personalities" with which their users could identify—truly something "mine" as differentiated from "yours."

The development explosion of recent decades also generated an unparalleled competition for land: land for housing, factories, schools, regional shopping centers, campgrounds, golf courses, athletic complexes, and an infinite number of other needs. In the search for acreage, vested interests militantly strove to outfox others in order to gain possession of the land they required. Others be damned. Too often this blind competition led to improper use. Facilities ended up on sites more suited for other uses, and those uses were being left with no appropriate sites. In the rush, sight was lost of the fact that land is limited and that its unbuilt-on reaches were diminishing rapidly. Leisure-service agencies were particularly affected. In countless cases, prime forest went under for industrial development while park agencies were left with naught but flat and barren fields.

In the late twentieth century, one million acres, an area approximately equivalent to that of the state of Rhode Island, were being bulldozed each year (Figure 1-6 on p. 8). Wise land-use judgments were critically needed, and land analysis strategies were devised to ensure that each interest could be satisfied, but only on parcels appropriate for its use.

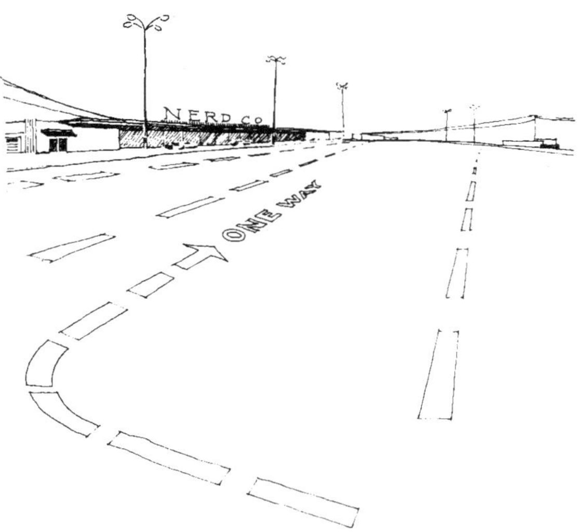

Figure 1-5
Bleak and barren and devoid of compassion for the human condition.

Facing the Present

In contemporary thinking, appropriate design is design that meets objectives considered particularly relevant for the individual park site. This is not to say that broad goals common to many park situations do not exist. Such requirements as good drainage and efficient air circulation must be satisfied in all solutions. What is implied is that design criteria should be developed through analysis of each situ-

Figure 1-6
"The good Lord is makin' more people, but he ain't makin' no more land."—Will Rogers

ation rather than through reflection on what has been found to be applicable to other circumstances. It follows that, even where some objectives repeat themselves often enough to be considered common to all developments, the manner in which they are best satisfied is most likely to be unique to each case. Each park has differences in site character at least. Hence, since what may be appropriate in one situation may not suffice for another, suspect any design solution which appears to be a rubber stamp etched from a prior scheme.

The unique requirements of a particular circumstance can only be discussed with information related to that job in front of us. Before dealing with such specifics, however, it would be useful to set forth those objectives which have been found to be common to most developments. To do so, let us first examine some conditions that portend to influence park design in this twenty-first century. By doing so, we merely acknowledge what history has taught. Recalling that the movement from geometric to nature-oriented design arose out of a desire to disassociate with the city, it may be concluded that meaningful work is done in response to the times.

Even though political climates change, environmental issues raised decades ago still demand our attention today. Too much of our present surroundings continues to breed tension, monotony, and anonymity. The production of visually refreshing, mentally stimulating, and pride-inducing parks as environmental exemplars is as important today as at any time in the past. Making wise land-use judgments is an equally valid contemporary goal. Land is a finite resource. Errors in assigning uses to land are a luxury of bygone times. Mistakes cannot be easily abandoned. Day by day it becomes clearer that there are no similar sites down the road.

It is also evident that today's work must be accomplished without the generous outlay of public funds that distinguished the immediate past. Inflationary times, cycles of recession, and changes in political attitudes have collectively created this condition. The immediate results have been a slowdown in land acquisition as well as a cutback in leisure-agency services. Public recreational needs have not diminished, however. How can those needs be met with static, oftentimes shrinking, budgets? That question is the fundamental challenge; the answer may possibly serve as a cornerstone for a more permanent and productive philosophy.

The first response to the challenge has been a search for sources to augment public moneys, since a basic level of funding must be maintained if anything is to be accomplished at all. For land acquisition, philanthropic foundations are being courted with increased fervor, and for development funds, joint private-public ventures are beginning to be discussed. Such operations, where private entrepreneurs might invest in the construction of profit-making enterprises (say, skateboarding complexes or aquatic playgrounds) on public lands, are argued by some to be the saving direction of the future. Indeed, a number of astute planning consultants, as part of their ordinary services, are starting to offer assistance in obtaining foundation grants and putting together funding packages with the private sector. Substantial attention is also being paid to the acquisition and recreational development of depleted mining sites, exhausted landfills, abandoned railroad rights-of-way, and other properties that no longer serve their original purposes (or are, for the owner, a liability, needing care but not producing a profit) and as such are often less expensive to obtain than untouched lands.

Heightened attention also must be given to increasing the productivity of the land agencies we already have. Multiple-use concepts are being explored with renewed vigor. These include ideas for multi-season use of spaces to broaden their active life (adjusting the summer baseball field for soccer in the fall). Another strategy has been to give recreational assignments to minimally used parts of the park which originally had exclusively a service function (flat rooftops have thus become gardens, sun decks, and temperature buffers). The challenge is to look at a familiar scene but see it in an unfamiliar way: from the height of a three-year-old; through the bifocals of an elder citizen; in three other seasons—wet, dry, snow-covered; filled with people celebrating a special ethnic occasion, eating, drinking, listening. What else could it be? (See Figure 1-7.)

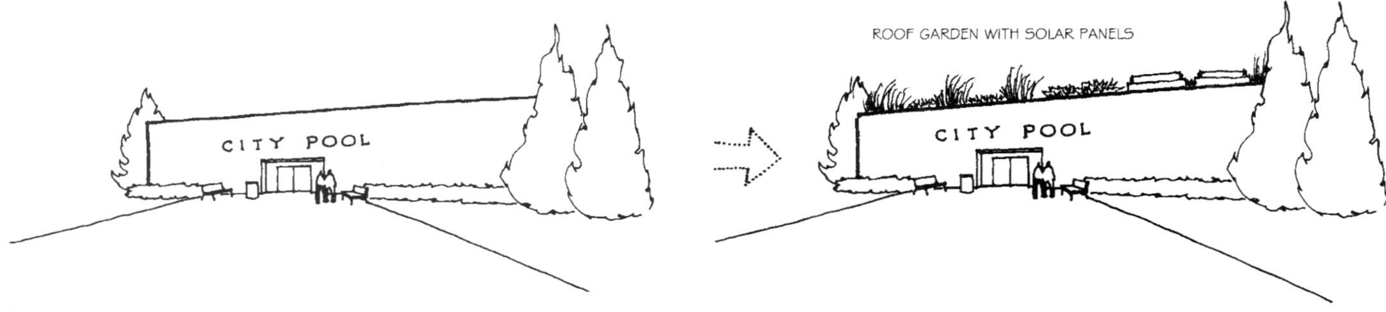

Figure 1-7
What else could it be? The City Pool building gains a roof garden and solar panels.

One of the most dramatic examples of "What else could it be?" is a park in an apparently impossible location, using a "useless" base for functions that seem to be expensively inappropriate. It's The High Line urban park in New York City. The High Line is a park is built on an historic, abandoned, elevated freight rail line above the streets on Manhattan's West Side. The City of New York owns it, and it's operated and maintained by Friends of the High Line Foundation.

The High Line is a mile long. Its breadth is the full width of the street below. Conceptually elegant and creatively witty, the High Line is accessible to the public all along its length. The park offers sun, seating, and opportunities to watch people and enjoy the elevated views of the nearby harbor. And New Yorkers love it.

"What else could it be?" At the extreme opposite end of the size and scale measure is the Great Park of Orange County, California. Formerly the Marine Air Station at El Toro, the site is, at 1,300 acres, about twice the size of New York's Central Park. Its size allows a staggering array of popular uses, including rides in a hot air balloon (which looks strikingly and appropriately like a gigantic orange). Both of these special parks have received national awards.

A trend toward the rehabilitation of old parks as opposed to the development of new sites entirely from scratch is also presently being witnessed. This is due in part to the funding crunch—fewer new sites are being purchased. But it also results from new activity demands of changing urban neighborhood populations and the pressures the new populations put on old places designed to serve bygone needs. As with other multiple-use questions, creative thinking about adapting original facilities is very much a part of the rehabilitation routine. The "flatten everything to the ground" urban renewal mentality of the 1950s cannot be afforded. Since it is essential to get the most from limited funds, the more that can be recycled into the new scheme, the better the plan.

The need to get the most from scarce dollars also gives new prominence to those traditional design objectives that affect energy consumption. Building structures oriented to reduce solar gain in the summer and exploit it in the winter, planning actions that reduce maintenance cost of outdoor areas—these are now more than agreeable goals. They are musts.

The contemporary mandate for increased productivity brings a new imperative for research that can lend greater certainty to design and management decision making. Information about the life span and usefulness of materials is especially needed. Surfaces: What materials last longest under various conditions of recreational impact? Plants: Which ones draw least on critical water supplies yet perform well functionally and aesthetically? Land itself: What are the least expensive, most effective materials and methods for stabilizing and upgrading sanitary landfills and similarly spoiled lands? Typically, answers to these and like questions have been pursued in academic or materials-industry settings. It is worth pondering whether results could be hastened with more direct involvement on the part of park agencies, which might put aside experimental areas in the parks themselves where tests could be run under actual conditions of use. Industry, as a potential partner in this quest, has much to gain in personnel services and much to offer in land and financial resources.

Findings from both general and specific research about people and their needs provide information that can make the outcome of planning and management actions more predictable, thus increasing productivity and gaining the best return possible from each dollar spent. To minimize errors in planning, broad assumptions about recreational usage are no longer enough. Responses from the specific populations to be served must be gathered and analyzed. Before the fact, surveys, formal or informal, must be made to determine what the future users of the park want to do, see, hear, enjoy. Recognizing the progress in speed and spread of information gathering, the opportunity to use electronic media for more and more focused information can be gathered using the technique called "crowd sourcing." New park idea? Posted on the World Wide Web, targeted to a specific population, thousands of focused opinions can be gathered for decision making in a matter of hours rather than weeks or months.

After the fact, observations must be made and objective conclusions drawn concerning successes and failures in user satisfaction. Are people doing what they said they wanted to do in the park? Are they sitting, playing, watching, picnicking where the design directed those activities?

The data must be gathered. The records need to be kept. The core of increased productivity is information—how much, how long, how many, when, where. These questions must be asked about whatever costs money, time, or commitment of assets. These records of performance, of both products and people, are vital to any conclusions drawn and directions taken. The problem with adequate data historically has been the sheer bulk of material that must be stored, accompanied by the difficulty of trying to evaluate all of it to successfully determine a direction.

But the handwriting is on the hard drive, so to speak (and as time passes and so does fading technology, it will be on CDs, DVDs, or in the "cloud," and in the future stored and presented through technology most of us can't even dream of). The computer hardware and software developments in the past have no apparent future limits. Computers have been integrated into every aspect of our lives and have already become the means of drawing conclusions from previously unfathomable masses of information. GPS (global positioning system), a technology as ubiquitous as the cell phone (or smart phone), makes possible the determination of a location, in three dimensions, within inches of accuracy of any place on the globe. Geographic information systems (GIS) have been refined to the same accuracy and make it possible to compile, store, retrieve, and analyze all kinds of information related to parklands and recreation programs. This digital tool is used to produce an amazing array of data in the form of charts, graphs, and maps. Likewise, with a similar facility, the Facility Mapping/Management (FM) tool can be used to inventory every aspect of parks from shelters to irrigation heads. This tool can track everything from last maintenance activity to facility usage levels, allowing an unprecedented level of park management. However, a computer is only a tool, not an end in itself. As a tool it carries with it the obligation of every user to use it well. Putting the best possible effort into filling information systems in order to properly feed the computer is justified by the computer's ability, unlike so many of its human partners, to remember everything it's told, cross-reference every scrap, and evaluate every alternative against all criteria.

Huge amounts of information are available and continuing to expand on the Internet, with which park designers and managers will be making better planning decisions. Agencies can now, with minimal investment in technology, keep all personnel—from top administration to the entry-level maintenance person—in immediate contact for communicating current information about improvements, maintenance, and facility scheduling. This capability comes at a time when park boards are finding value in minimum investment and maximum productivity, and are recognizing these tools as a means to that end.

Contemporary Design Goals

Every action begins with the identification of objectives. While no two design projects are ever alike, each having differences of site, facilities, users, and other parameters, there are several goals that remain constant for all projects.

In the most general sense, foremost is providing facilities that lend grace to the environment, provide mental stimulation, have pride-inducing personality, and make best use of the land. A commanding priority, however, is implementing planning measures that increase the productivity of sites put to recreational use. These are traditional design goals, but they are singled out for special focus here in response to the unique demands of our times and our foundation-solid sense that these concerns should be part of any permanent philosophy.

Within this contemporary context, we can identify further common objectives to be met in the design of park properties. In the following chapters, these are offered as principles or broad guideline categories in which judgments must be made as the designer proceeds through the problem-solving process. The principles are equally applicable to sites for passive uses and those intended for active functions. They are addressed to the satisfaction of general contemporary needs as well as to the specifics of individual projects.

These principles are then broken down into subparts, called *matters of concern*. Each concern should be weighed when considering the variables of any project.

Principles and matters of concern are the *whats*. Illustrations are laced throughout to indicate a few *hows*, ways in which principles might be satisfied and ways in which the matters of concern might be treated. Since it is impossible to set down every possible circumstance, illustrations are meant merely to indicate the wide range of opportunities for principle satisfaction. In addition, the appropriateness of the *hows* must be considered on the merits of the cases in which they are presented. Quite reasonably, what satisfies for one instance might not work for another.

2 The Umbrella Considerations

This first set of considerations can be read as a broad overview of the principles that guide design decisions. It comprises the overarching framework to which all other goals must relate. The three principles to be discussed in this framework are that everything must have a purpose, design is for people, and both functional and aesthetic requirements must be met.

Principle 1: Everything Must Have a Purpose

If wise land use is essential, there is no room in site design for whimsical judgments, even though many design ideas may come to the designer intuitively. Intuition is a means or a mental shortcut to solution, but it is not a justification for results. Decisions can and should have convincing backup. They must be supported by sound and logical reasons.

Design elements, therefore, must have an identifiable purpose. One such purpose is to establish appropriate relationships between the various parts of the park complex. These parts include *natural elements* (for example, land, water, plants); *use areas* (game courts, ball diamonds, parking lots, roads, walks, maintenance yards); *major structures* (buildings, dams); *minor structures* (drainage, electrical and other utilities, fences, benches, drinking fountains, signs); *people*; and other *animals*. In addition, all are affected by *forces of nature* (wind, sunlight, precipitation). (See Figure 2-1 on p. 14.)

While each part will present its singular demands, no part can work in isolation from another. Follow steps A–F in Figure 2-2 on p. 15 and note that, for instance, orientation to the sun affects the location of the amphitheater, for the sun must not shine directly in the eyes of the audience (A). The amphitheater siting directs the placement of the parking lot (B), which narrows down the possibilities for

Figure 2-1
A park is a complex of many parts.

access routes to the public streets (C). The amphitheater parking lot could serve the nearby marina, but the marina noise, in turn, must be neutralized before it reaches the amphitheater (D). The traffic invited to this amphitheater-marina complex must not conflict with surrounding land uses, as it would if it proved a safety hazard for children crossing neighborhood streets (E). On the other hand, the smells, noise, and visuals emanating from surrounding facilities must not cause the enjoyment of theatergoing or boating to be lessened (F).

Interdependence among all the parts must be recognized and accommodated if any single part is to work. Consideration of such relationships extends from the broadest determination of the park's place in the city plan to the smallest decision about where to place the trash basket.

Matters of Concern

1A. Relation of Park to Surroundings

Design focus must extend beyond the park's boundaries in order to answer such questions as: Will the proposed park development cause flooding in the valley below? Will it cause traffic to back up into residential streets? Will the arrangement of the new facilities replace the pleasant view of undeveloped land with an unpalatable distraction? Will the location of play areas encourage baseballs to fly into backyards?

The reverse, the impact of the surroundings on the park, is equally significant. Will adjacent land uses send stomach-churning odors across proposed picnic areas, create safety hazards, or destroy the park's serenity? In a positive sense, do the surroundings offer potentials to be "borrowed"—a good view, a major access route, a utility source?

Through purposeful design measures, a sharp designer exploits the advantages of surroundings and overcomes the limitations posed by adjacent lands and their uses. The designer also attempts to see that the park remains the best of neighbors, never causing the usability of surroundings to be diminished.

1B. Relation of Use Areas to Site

Land must not be wasted. Every corner of every site must be assigned a use. This does not necessarily mean active use. Fallow land may serve as a buffer or a viewing panorama. It may be left as conservation acreage if it has been determined that active trespass would cause loss of wildlife cover or vegetation that holds back floodwaters. Even an undisturbed swamp serves a valuable purpose, for its porous surface allows rainwater to penetrate and refill underground reservoirs. Fill in or pave over the swamp, and the water supply for neighboring habitations diminishes.

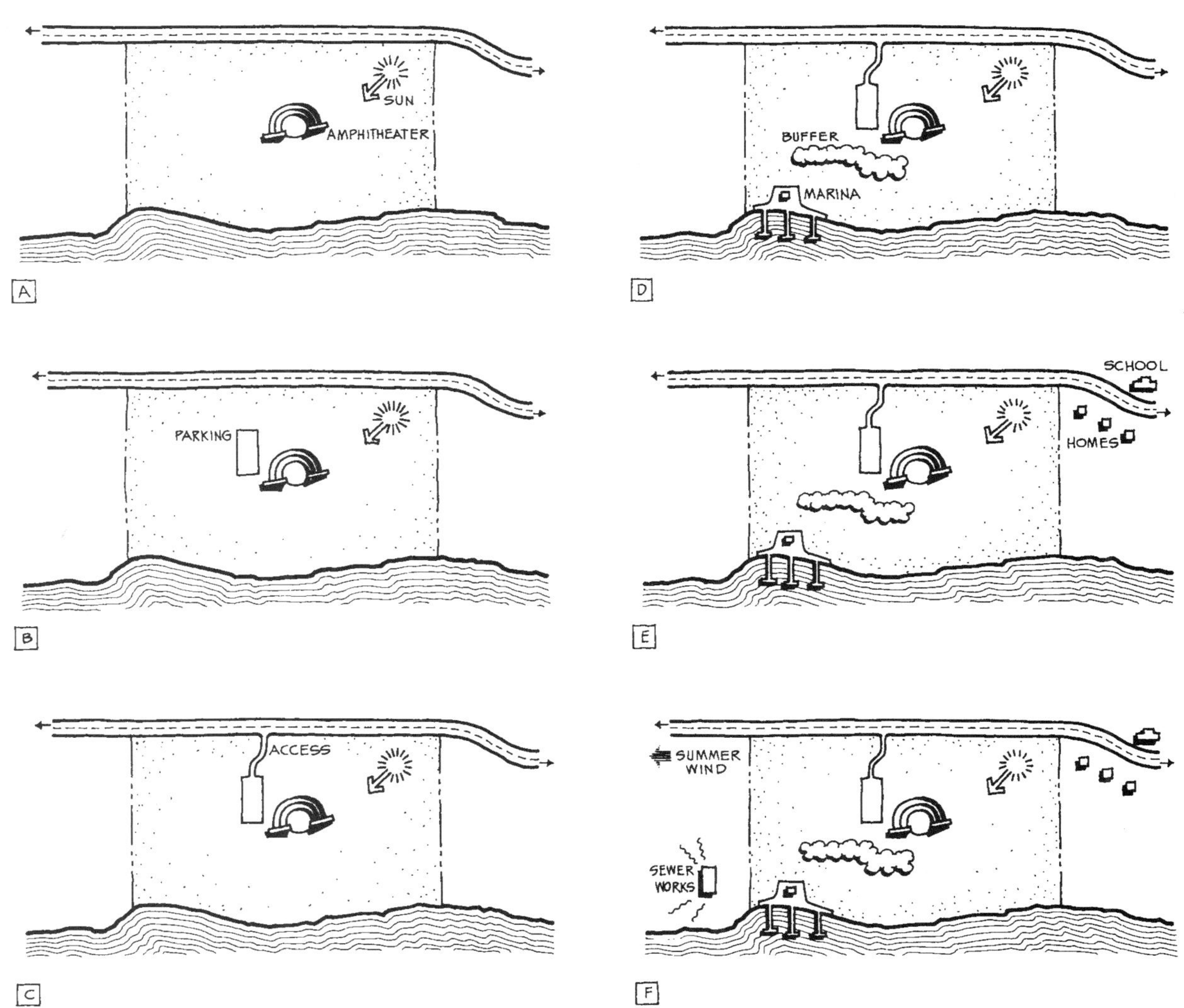

Figure 2-2
The location of every park unit affects the workability of another.

Figure 2-3
Use areas may require different degrees of slope.

Whether they are for active or passive uses, *facilities should be assigned only to portions of the site that are compatible with those uses*. As an example, consider the implications of a single site factor: degree, or steepness, of slope. As illustrated in Figure 2-3, tennis courts cannot function unless they are on flat land. Tobogganing must be matched to relatively steep slopes. While neither tennis nor tobogganing do well on precipitous slopes, these grades are ideal for separating two nonconforming activities, such as a gathering for a church youth group and a fraternity party.

All uses can be cataloged as to their demands for slope and much more: type of soil (stability, fertility, permeability); need for vegetative cover; nearness to water, utilities, and transportation; and orientation to sun and wind, to name a few.

Therefore, it is essential in design to identify both the limitations of the site and its potential, to overcome the former through the thoughtful location of facilities and exploit the latter. To bring this point home, see Figure 2-4 and consider the consequences if use areas are not located where site characteristics are compatible: Picnicking is designated for the sun-baked field, while nearby trees are cleared for the parking lot; buildings are constructed on shifting soil bases, while intermittent walking routes are laid out on adjacent stable surfaces; the amphitheater is situated in a howling wind tunnel, the winter roads where the snow loads are heaviest, and the ball fields in the marsh.

1C. Relation of Use Areas to Use Areas

Before being assigned to locations on the site, various uses should be analyzed in terms of compatibility with each other. Such an analysis unearths both common and disparate threads among the units. For instance, as illustrated in Figure 2-5, nature walks, canoe lagoons, and sitting areas can be considered alike for they are quiet and soul-satisfying; tennis courts, handball surfaces, and basketball pavements are related because they are noisy and sweat producing and are therefore opposed to the quiet grouping.

By locating common units together and segregating them from incompatible use areas, activity enjoyment is enhanced; such distractive battles as noise versus quiet are eliminated. In addition, movement patterns are simplified. Those with an interest in one type of activity may choose to proceed to only one area to find a full range of desired facilities. Area supervision is also made easier. A single supervisor can keep track of, say, preschoolers if all of their activity areas are grouped together, whereas more personnel would be required if such play zones were scattered all over the site. And maintenance procedures are simplified. Since similar activities usually require similar maintenance chores, performance time can be minimized if mainte-

Figure 2-4
The results when use areas are located on unsuitable portions of the site.

nance equipment—trash trucks, horticultural equipment, infield draggers—are able to concentrate on clustered areas.

In weighing similarities among use areas, sometimes clear-cut decisions can be made. Most often, however, use areas are found to be interdependent for one reason, but incompatible for another. For instance, maintenance yards conflict visually with picnic sites, but movement between them must be easy; parking-lot noise detracts from the concert, but the lot must be located nearby; playground complexes are for kids but, for supervisory purposes, must be associated with adult areas. However conflicting, all such demands must be treated in the use-area organization proposal. Perhaps in the latter example, the playground could be physically separated from the adult station by a low barrier, yet remain visually evident—over the barrier—for supervision.

After relationships between use areas are determined, the design task becomes one of finding sites that fit the desired pattern; for example, for picnicking a well-drained sector with ample shade trees and stable soil lying next to a cleared patch suitable for parking but away from the school building situated just outside the park's limits. (See Figure 2-6 on p. 18.) Thereby an "ideal" relationship is struck in accordance with all three matters of concern discussed so far: relation of use areas to use areas, use areas to site, park to surroundings.

1D. Relation of Major Structures to Use Areas

In a sense, a building may be thought of as a use area and decisions regarding its location made on that basis. Special attention should be paid to the relationship of various rooms to adjacent outdoor areas. This will raise such questions as: Is the gym entrance handy to the playfield? Can the children move from the kindergarten to the "tot lot" without having to cross the parking area? Are classrooms buffered from noisy game facilities? Can nonswimmers move from the bathhouse to the wading area without having to walk along the edge of the deep pool?

1E. Relation of Minor Structures to Minor Structures

Just as a park is a complex of related areas, each use area is a complex of interdependent physical elements. As secondary as these concerns might be when weighted against the other relationship questions, inattention creates rightful public irritation. Is the bench close to the refreshment stand? Or must a parent juggle a half-dozen ice cream cones for 100 yards before he or she is able to sit down amongst the brood? Is the seat near enough to the light fixture that is intended to illuminate its surface? Is it the right distance from the trash basket, thereby accommodating the easy flip of the discarded newspaper? If the basket is in an inconvenient location, the paper is going to stay on the bench waiting for the wind to blow it all over the site.

Figure 2-5
Compatible uses should be located together and should be separated from groups of disparate activities.

The Key Word

Establishing ideal relationships is only one purpose of design. Other purposes will be articulated later. For now, consider that this introduction to relationships has been set forth to illustrate the broader point that every design decision must be made for a logical reason. To ferret out purposes that may be at odds with your own and to guard against arbitrary judgments and half-baked conclusions, the simple question to ask of a designer is: "Why? What is your purpose?"

Figure 2-6
An "ideal" use relationship.

From the broadest concern ("Does this facility belong in this part of town?") to the most incidental detail ("What underlies the selection of that material, color, shape, height, width?"), expect "Why?" to be answered to your fullest satisfaction.

Principle 2: Design Must Be for People

People are the benefactors of any park development. However, development success is often measured entirely by how well it meets the demands of machines and equipment, how it simplifies administrative paperwork, and how it fits an unyielding formula of quantitative standards. Perhaps it is because some minds have become immune to the knowledge that these impersonal requirements are merely means to serve people, not ends in themselves.

The elevation of impersonal requirements to prime measures of development success can lead to the creation of an uncomfortable mold into which people must then be forced. This conflicts with design purpose, which in park design is to develop an environment that fits people.

It may be most helpful to begin the section by describing the various people (individuals or groups) who may be involved with the design, construction, maintenance, funding, and use of parks. Individuals within this scheme may play more than one role.

Figure 2-7 is a diagram showing the relationships between people in the anatomy of a park. Part of the distinction of individuals' roles may be whether the person is serving or being served. The designer is serving the park board in coordinating the park design process, whereas the park board is serving the public by making the tough budgetary decisions about park construction and maintenance.

Park Designer/Landscape Architect. This person or team has training in design and is going to coordinate the design process. While he or she is hired by the park board or committee, the responsibility for appropriate and creative park design is primarily directed at the park users. This role may be played by a consultant or staff person. The park designer must listen diligently to all of the other people in their various park roles.

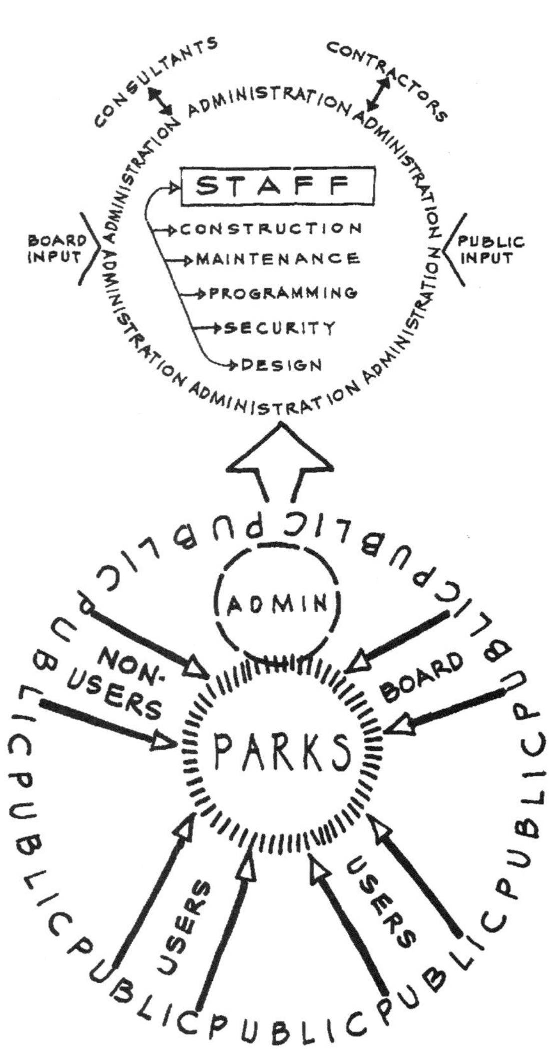

Figure 2-7
Anatomy of a Park: Relationships.

Park Board or Commission. This group of individuals is a volunteer group, appointed by the mayor, town manager, or town board to administer the parks (and recreation) department. It makes decisions about park construction, maintenance, and programming, and it allocates money from its budget. The park board hires and dismisses park employers or designates that task to the park director or superintendent. The park board is the fiscal agent. They make plans and set priorities. Their concerns must be heard.

Park Staff. These people focus on various aspects of park development and programming. The park superintendent or director assigns and oversees all staff activities. The size and complexity of the park system and parks budget determine the number of park employees. The superintendent may be key to the park design process, but his or her opinion is not the only valid set of concerns.

Park staff may be assigned to plan and design new parks or to improve existing parks. Some of these projects may require outside consultants. The staff may be capable of doing the design but may also be of great assistance to the design consultant, if one is employed.

Although some park staff may be assigned to construction tasks for new parks or renovations, it is more common for new construction to be done by an outside contractor. Contractors or construction staff have experience that is invaluable to the designer. They may have great insight into the construction process, and they tend to know what works.

Maintenance staff are responsible for the upkeep of equipment, playfields, natural areas, and all facilities. Most maintenance tasks are done in-house, but some are done by outside contractors. Maintenance staff are quite attuned to the maintenance needs of various design elements and can offer ideas to reduce maintenance issues of new park designs.

Programming staff are responsible for organized activities such as soccer leagues, summer recreation offerings, and nature interpretation. Program staff often rely heavily on volunteers to help conduct the program required or requested by park users. They, too, can help the designer assess current and projected park activity needs.

Park security may involve yet more staff. Security is usually handled by the local police force, but large park systems may have their own security personnel. Regardless of who provides the service, the designer must be aware of security issues and resolve not to create new ones.

All these people are, in different ways, in roles of service to the public. Administration, design, construction, maintenance, programming, and security activities all focus on the provision of enjoyable, convenient, and safe parks for people. Each has some information about how to provide the best parks for the available money. At this point, a worthwhile step for the director to consider would be an informational meeting (more likely a series of meetings) to educate all the groups identified above regarding the functions and *values* of each group to all the others. Groups with specialties also tend to have language specialties (jargon). The most useful service the director can bring to the process of this education is being the translator ("What it means in 'People Talk'"). A consultant/outsider should be included to complete the perspective. It's a good role for a valuable and interested guest to provide. Likewise, a public representative should be included (board members would be appropriate) to listen, learn, and spread the knowledge and appreciation gained. This is definitely not a one-time offering. It should be a continuing effort, since so many changes happen so fast in every park system, small or large.

The other side of this story is about the public at large, and more importantly, the park users. This group is discussed at length in the remainder of this chapter. Even though most people use parks occasionally, some are frequent users and others do not use the parks at all. The frequent park users are going to be the most vocal in expressing their needs in the planning process. (Nonusers may be vocal about excessive spending on "frivolous" park features. They may think the money should be spent on city utilities or other public projects, or that the money should be saved to reduce taxes.)

Matter of Concern

2A. Balance of Impersonal and Personal Needs

While it is essential to meet the requirements of machinery, that is not enough. The demands of vehicular traffic, the gang mower, and the utility network must not overshadow the needs of the people whom these inanimate objects are meant to serve. Both sets of requirements must be given full consideration. For instance, in the design of roads and parking lots, it must be recognized that the automobile demands a certain road alignment, gradient, roadbed structure, and curb height, as well as a 9- by 20-foot paved slab for storage. It must be remembered at the same time, however, that the people inside the vehicle seek visual refreshment and mental exercise. As exemplified in Figure 2-8, these could be provided for them by roads rhythmically curving between softly rounded earth shoulders, revealing appetizing views, with peripheral distractions screened and oppressive natural elements shaded out—as the motorists wheel over proper alignment, up and down appropriate gradient, between suitable curb heights, and into the parking lot.

Administrative efficiency is likewise an essential in park agency operation, and it is often met through the purchase of standardized equipment. Certainly, it is simpler to fill out an order blank for 100 standardized items than to requisition the same number of custom-tailored units, and it is economical to purchase mass-produced equipment. But what happens when the result is that everything looks alike, looks alike, looks alike? Mental stimulation is suppressed. Individual identity is smothered in anonymity.

Can the economic benefits of standardization and mass production be realized without cheating the prime benefactor—the user? The designer should consider dovetailing the previous principle into this one by applying the key word, *why*. Ask: "What is its purpose?" If park benches are primarily for sitting, they might reasonably be mass pro-

Figure 2-8
Roadway design should meet both the demands of the vehicle and the needs of the people inside.

duced if they prove comfortable, and look-alike items distributed about the park if the visual effect of their sameness is submerged in an overall uniqueness about the development. This is another reason why rubber-stamp layouts should be avoided; if the design is unique, the standardized parts of the scheme will be only incidental portions of a refreshingly individual whole. A park designer will be able to demonstrate a design vocabulary through the artful selection of site accessories.

In addition, for a purpose such as the communication of information, standardization can actually benefit people. Since repetition creates familiarity, similar styles for all signs and waste receptacles, for example, provide immediate signals for users in the time of need. Because one has seen the object previously and identified it with a particular function, one knows what to look for. This familiarity helps the user to spy it quickly (Figure 2-9).

In contrast, "typically" standardized playground apparatus, with the same swings, the same slides, fails when measured against purpose. As a major unit in the park, the play complex is likely to have significant visual impact. Sameness here can lead to dulling visual appetites, especially among the adults who are there to supervise their kids or who pass the area every day. The sameness of traditional play apparatus may actually frustrate one of the purposes of play. Such stock items can only be used in one fashion and therefore stifle a child's discovery mechanisms. This deficit is compounded by the fact that even from the start, the kids have no say in the way in which each piece can be used, for a single way has been dictated by the item's design. Conversely, in an attempt to use play equipment creatively, kids may actually finds unsafe ways to explore their limits.

Awareness of this need for variety and liveliness in total design has resulted in the better equipment manufacturers devoting time and money to the development of creative, safe, and exciting play equipment. Their marketing includes offering site-design services or consultation with the site designer to integrate the equipment into a park's total framework. The results are good, showing a perceptive appreciation of color and form as well as sensitivity to the playability of the area. Probably the most important factor in the presence of the equipment company in these scenarios is the legal implication of liability on the part of the designer of a playground for injuries resulting from the use of equipment in the area. The manufacturers take the initiative to research and test their products, and are consequently able to take the legal responsibility for litigation arising from accidents.

Yet, as a result of unmindful inertia, standardized equipment (traditional playground apparatus is a classic example) continues to be purchased long after it has outlived its reason for being. Coincidentally, the question of this equipment's usefulness brings us back to the question of purpose—which at its root is still this question: Why? In this case: Why do children play? The comfortable old theory came with the primary title of *surplus energy* (kids have these poisonous humors within that get pumped out only by working like crazy on equipment type X), followed by the subheadings *motor exercise needs* (in which the upper torso, lower back, upper arms, and other body parts must have developmental support through specific activity on equipment type Y), and *social skills* (the innate craving for social order and equity is brought out and polished by the opportunity of taking turns waiting to use equipment type Z).

That's the theory. Sorry. Turns out that play is . . . *just for fun*. In Mike Ellis's book *Why People Play*, his more specific answer is that

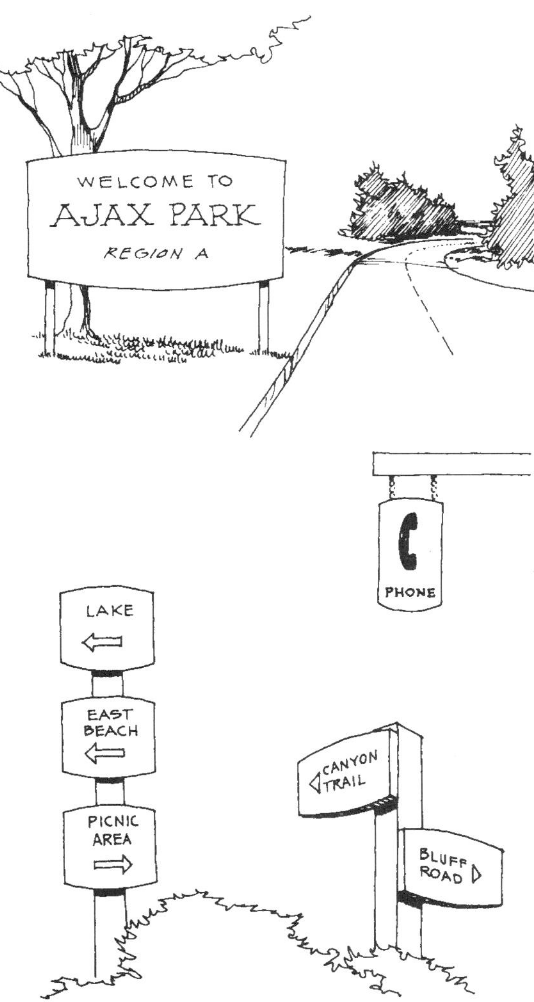

Figure 2-9
Similarity among sign styles helps the traveler in search of a message.

they play *for the stimulation they receive*. In *A Visual Approach to Park Design* Al Rutledge points out that after actually *observing* what was going on in a playground, without prejudice of a filtering theory, it became obvious to him that the kids weren't staying on *any* equipment long enough to burn energy, develop motor skills, or learn the kinds of social skills touted by the advertisements. Their target was purely the momentary stimulation offered by the equipment—one shot and on to the next piece. Yet this discredited theory is the basis for the play-equipment standard.

A discussion of standardization in park planning necessarily reaches further to consider use of the popular tables that propose quantitative standards as well as types of facilities and activities for typical park properties (see the sampling included in Appendix 1). Since full lists of these quantity standards, which express national or regional averages, are found in almost every book dealing with recreation area development (as well as in many design-data manuals such as Harris and Dines's *Time-Saver Standards for Landscape Architects*), there is no need to present a complete tabulation here. However, it is relevant to this discussion to forewarn about their blanket application, since to apply these standards uniformly to all properties is, at best, to depersonalize development.

A decision on the degree of attention to be given to any quantity standard should hinge upon some understanding of how the standard came into being. We have gone through that process with the play-facility standard. It evolved from an anachronistic theory. It deserves to be rejected out of hand. The play-facility standard also implies that children from tots to teens have the same play needs—a curious implication. This example raises the most fundamental question that must be asked of any model: To what degree does it suit the specific and varying demands of those who would use the works? As an example, let's take a serious look at standards for activity areas, specifically the widely printed suggestion that every neighborhood park should contain a softball diamond, playground equipment, multiple-use pavement, and a turf area. As with most standards, this is based on national averages without any allusion to the user groups whose needs the recommendation pretends to satisfy. Even where it could be argued that it reflects a majority trend, its usefulness as a planning guideline is marginal since in a given place any national minority can be the resident majority (Figure 2-10). And it is the specific place with which a designer must deal.

Figure 2-10
The resident majority.

Sadly enough, no matter how emphatically standards might be labeled flexible, the temptation to regard them as absolute is great. They appear to offer a substitute for thought and an easy path to solutions. *Facility and activity standards must be taken simply as points of departure that deserve to be modified according to the unique demands of each and every circumstance*. Clearly, parks for the economically deprived may require different elements than parks serving the more secure middle class. Surely, conclusions about an urban neighborhood are not equally valid in a rural village. And likewise, parks in neighborhoods with substantial teen populations should have facilities that are different from sites in neighborhoods overflowing with tots or unmarried young adults or senior citizens.

Obviously, any development responsive to people provides activities and facilities tailored to the clientele at hand. Ideas for these are more likely to come from recreators with a finger on the pulse of the locale and on-the-spot research than from impersonal generalists far removed from the scene drawing up lists in their remote offices. Suffice it to say that gathering and analyzing data on leisure-time needs, as expressed by the potential park users themselves, is of primary importance. These *demand studies* are typically accomplished through questionnaires, interviews, and public meetings called when a park is due for capital improvements.

But while demand studies do help bring the local people into the development process, a nagging worry remains; demand studies do not reveal every facet of need. Whatever people say they want springs from a need and should therefore receive substantial attention. However, it is rare for people to possess enough insight to identify every condition that could benefit them. Accordingly, it is incumbent upon professionals to take up where public articulation falters, folding into their work human satisfaction factors seldom expressed in public proclamations. Realistically, the science of sorting out little-understood human needs is inexact at best. Its application is often checked by the ever-present fear of becoming a self-righteous moralizer rather than a servicing prophet by mistaking a purely subjective conclusion for one that indeed does contribute to the common good. However, questions raised by physiologists, psychologists, sociologists, anthropologists, ethnologists, and others attempting to understand human behavior point to areas in which leisure-time experts can operate, translating reasonable hypotheses and theories into activity programs. The ideal effort is a collaborative interplay among researchers and appliers.

Park users themselves are a primary source of information on leisure-time needs. However, they may be unable or unwilling to articulate these needs under direct questioning. *A Visual Approach to Park Design* demonstrated how both planners and recreators might unearth design-relevant insights through the simple process of casual but sustained observation of parks in use. Those insights can lead to generalizable ideas about how to produce satisfying environments, as in the case of the designer who noted children placing planks between play devices and enthusiastically crawling, jumping, swinging, and wrapping themselves around them while continuously moving from one piece to another, innovating all the way. From this he concluded that for optimum play fulfillment, all playground pieces should be connected rather than spotted about as was (and continues to be in many places) the most common practice (see figures 2-11 and 2-12).

Figure 2-11
Organizing play pieces to provide experience continuity conforms to an observable play pattern among children.

This attractive hypothesis led the designer into a sequence that demonstrates the concept of park as design laboratory. Before creating a new playground, the designer gathered data by observing existing play facilities and objectively evaluating the observed activity as real expressions of need by the children (rather than "incorrect" behavior because it wasn't what would have been expected). The speculative conclusion drawn from the observation then became the basis for a physical design experiment testing satisfaction of those needs. Continuing observations *after* the fact provided the confirmation of the speculation: Observed needs were real needs, satisfied as play followed the patterns predicted by the design, supported rather than impeded by the equipment.

Figure 2-12
Where playground devices are isolated from one another, the potential for continuous play experiences is diminished.

Observation alone may not enable a designer to identify the needs of special populations. Where the population of "watchees" differs significantly from the "watcher," all action may cease when deliberate observation begins. The target population either may disappear or become observers of the observer. Confronted with this situation, some designers may reduce their observation to a minimum and

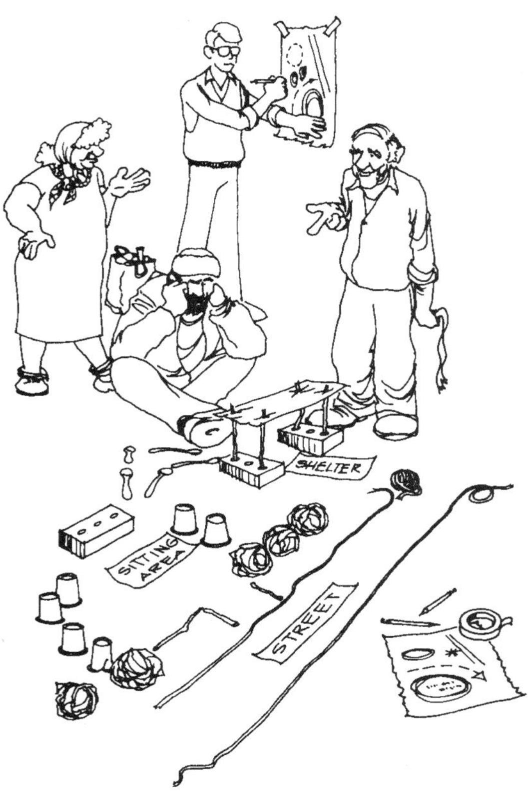

Figure 2-13
Planning sessions with a special clientele require special preparation.

increase their speculation to a maximum. If they see that vagrants disappear once a park has been cleared to permit viewing from the street, they might speculate that the most comfortable environment for the vagrant is one that is screened from the world of the affluent. The more likely reason for disappearance of the vagrant population would be the loss of the comfortably separated setting for watching the passing parade rather than the need for screening. Who is screened from whom? But this remains speculation without local site-specific research.

Recognizing the hazard in this kind of speculation, a designer in California determined to go beyond the observation stage and cursory laundry list of ideas when he developed a very successful slum-edge park. The clientele, definitely a special population, were actually participants in planning sessions concerning the design program for a park space in their territory (Figure 2-13). With three-dimensional apparatus as basic as paper cups and sticks, ideas were exchanged and tested and needs were identified; through the participation an interest was achieved that has sustained the constructed park as an environment that the users enjoy and that the city can tolerate.

The designer acknowledged that the first level of participation would never have been achieved without some incentive. In this situation it was the offer of $5 and a free meal—*after* the planning session. Special populations may justify special methods as in this case. The wisdom of the method is confirmed by the after-the-fact support of the park by its special population.

Interestingly, in a similar setting with a similar population in another major city, a park design built on assumptions based on observations affected by considerable (unfortunately unrecognized) designer bias and completed without client participation was voted a loser by the local barrio population. They proceeded to disassemble it splinter by splinter in a matter of weeks. Failing the after-the-fact test, the park is confirmed as not acknowledging the needs of its special population. Neither the designer nor the agency can afford practice shots. Observation interpreted to reveal real needs and augmented by participation of a special population is design insurance and a designer's responsibility.

In other economic and cultural settings—for instance, upper-middle-income singles in condominiums or senior citizens in their retirement enclaves—participation may be just as difficult to achieve. Residents may say, "Good heavens, we can't decide—the designer's the professional," placing the same kind of reliance on the designer's expert opinion as on a lawyer's or a doctor's. In this case the skill of the observer, effort devoted to observation, and background knowledge of parallel situations are critically important. The designer must put what little is said together with much that is seen and derive needs that are valid. The client vote on a completed project in this situation may not be as violent to the physical setting as in the barrio arena (site of the splinter-by-splinter dismantling). In fact, as a negative, it may simply be the "no-touch-no-talk" treatment. The damage that results, which is professionally more deadly, is to the reputation of the designer.

In order to bridge the dangerous chasm between park as plan and park as product, the designer, knowing that users may have difficulty interpreting the plan drawings, needs to find practical ways to help them "get their minds around" the plan. The designer might use the California designer's interactive method or employ sketches, built models, or computer simulations. User representatives might visit other parks with similar facilities that can serve as demonstration labs for their park development. The bridging, however accomplished, is very important. Communication is the key.

The perception of the park as laboratory offers designers and recreators a potential so far only partially tapped for evaluating the overall success of design work. Besides unveiling hidden aspects of recreational need, observation of newly constructed parks in use can also reveal much about how well the planning dollars have been spent. This type of evaluation requires that all design objectives for a project be treated as hypotheses or hunches about how people will use or otherwise react to specified aspects of the work. Post-construction evaluation findings provide evidence of what worked and what didn't. And such evaluations go a long way toward pinpointing where adjustments are deserved or preventing the duplication of mistakes in the next job.

Sadly, post-construction evaluations of park properties are rarely if ever conducted. Conceded, the process imposes an additional burden on the staff. Yet that seems a shortsighted argument when the purpose of the test is the proof of maximum productivity per dollar spent. Steps to ensure that design work gives the highest return in user satisfaction can ill afford to be treated as niceties. They are necessities. To heighten human-satisfaction returns, we must also pay closer attention to what the social and behavioral sciences have to say about human needs and incorporate the best of those offerings more thoroughly and consistently into the planning of physical works.

The often-expressed hope of designers for factual support of behavioral theory is being met as more useful publications are hitting the stands. William H. Whyte's now classic studies of people in urban spaces are an example of the knitting of threads of theory into the fabric of fact; Whyte's theories are supported by documentation of people observation in plazas, squares, mini- and maxi-parks, and all types of urban areas where leisure time is spent. The appearance of his *The Social Life of Small Urban Spaces* as book and subsequent film was an event of critical importance to anyone involved in urban-area recreation design. Whyte's work includes examples of techniques used in gathering data from and about human subjects, who inevitably seem to be ham exaggerationists in the presence of movie or video equipment. Whyte's techniques as discussed in the book include simple people-mapping procedures as well as more elaborate but very effective time-lapse filmmaking techniques.

A Visual Approach to Park Design is another useful reference. It conducts a search-and-distill mission through the rapidly expanding literature concerning people observation. It presents a compendium of the knowledge of territoriality, personal space, and interactive social behavior related to park and recreation activities and explores the design implications. The essence of the literature, which should hardly be surprising, is again the critical importance of observing people to understand their needs. An important emphasis is on how to train one's own eyes to really *see* what one looks at. Many data are offered, but the reader is cautioned that behind any generalization is the *exception*. The particular issue or characteristic that may necessitate direct observation and knowledge of that situation may include the following: the physical setting, the reaction of the people to that setting, the people in the setting of their own current lifestyles, the people as seen by themselves, what they think they are, what they like, and what they do. People watching becomes the central and critical requirement for

Figure 2-14
Orientation of sitting places toward activity spots encourages human interaction.

Figure 2-15
Grouping of benches fosters conversation and silent inspection of others.

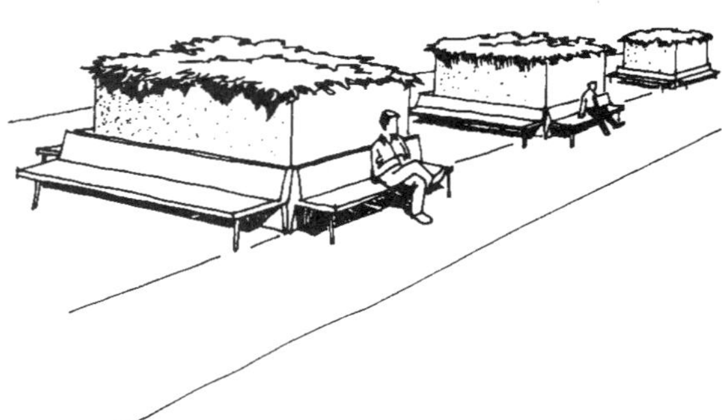

Figure 2-16
Isolation of benches hinders being with others.

the designer, but—*and this may be surprising to some designers*—it is also the most desirable *activity* for people: seeing and being seen. This is a need that includes and transcends all other activities and encompasses all categories of people regardless of age, sex, and interests.

Consequently, activities *plus their settings* get to be successful people-pleasing elements very much to the extent that they permit and promote the visibility of the action; no matter that it ranges from almost passive non-action like talking to extremely hectic super-action like a pickup game of basketball (Figure 2-14).

The "two's company" of interaction and meeting people is another all-ages need basic to society; it ranks just below seeing and being seen. Where interaction is to be encouraged, benches should be grouped to face each other (Figure 2-15), never placed back to back or in isolation. In addition, distances between benches are important to conversation making, as indicated by Robert T. Hall in his book *The Silent Language*. Anthropologist Hall points out that members of different cultures automatically assume unique distances when conversing, and he lists what he has found to be the most comfortable talking distances for Americans. In his book *Personal Space: The Behavioral Basis of Design*, psychologist Robert Sommer reports that back-to-back and far-apart placement of seats is a technique used in airline terminals to drive people from the waiting areas into bars and coffee shops where the atmosphere is more conducive to conversation. (Not coincidentally, this is where the terminal can make a buck as well.) Therefore, if you see park benches arranged like those in transportation terminals—and most of them are—you may conclude that the designers wanted to discourage interaction, or drive visitors into bars, or didn't know what they were doing when they ordered the benches set out (Figure 2-16).

Freedom to be unencumbered by domineering authority and chart one's own course has been identified by some behaviorists as another human need. As we will see in the section dealing with circulation, landscape architects can meet this requirement by organizing facilities and traffic-ways so as to guide visitors into use patterns which they might have chosen for themselves, rather than force them to move about only as others demand. *Freedom* also suggests a need for space that is not allocated to predetermined activities. As exemplified in Figure 2-17, these would be arenas for satisfying whim, "do as you please" areas, places to throw a ball, lie on the grass, chase a greased pig, walk the dog, or do whatever suits your fancy at the moment.

The designer's understanding of these needs and others unique to a population is distilled from the process of looking and listening. Getting to know the user is the most important step toward the goal of creating a design to satisfy that user. No formula is offered or available for satisfying these needs. Realizing that the needs can be identified and that the means for doing so is improved and expanded people watching doesn't make the process of satisfying them simpler. But at least designers should not ignore clues thrown down in front of them. In a conversation, a landscape architect marveled at the panoply of bright colors displayed in the dress and other physical symbols of an ethnic gathering. Yet, the brightest color found in a major renewal development he designed for their neighborhood was a tepid grey. The designer admitted two things: The people identified with bright colors; they hated the new development.

The burden returns to the back of the designer. The people contact has to be made, and its depth and duration is a direct correlative to success in learning their needs and creating a good design for them.

Figure 2-17
Simple open space should be set aside to satisfy the leisure-time whim of the moment.

Spin-offs like improved security, reduced vandalism, and user satisfaction are the frequent and sometimes unexpected coincident results.

At the risk of allowing reemphasis to become overemphasis (a risk worth taking in this case), two professional pitfalls must be rigorously avoided.

The first, the rubber-stamp replication of standard facilities, entails danger because it transfers a conclusion from one source to another setting. Each situation must be met on its own merits. Standards are only starting points.

The second, the imposition of a personal view, no matter how well intended, is a destructive disservice if it crushes what the designer should regard as a blossom—the user's need. The parallel to a tender plant is appropriate. The need expressed may be an unrecognizable, even feeble sprout, needing the water and care of a creative design response, not to be choked out by a weed-like standard that "will work because it's worked everyplace else." Under the pressure of budget and time constraints plus users' admitted dearth of knowledge of their own needs, the designer is still obligated to remain objective and nurture the needs expressed or discovered in the park-laboratory. Before time and money are spent constructing a hunch, observations made with eyes wide open will definitely be informative and probably fun to make. To look at people, before and after they have a park, and *see* what they do is an effort that must be made.

The issue is that of professional responsibility. The professional with a balanced view of responsibility to and for the client-user will focus on the creation of order from the apparently chaotic, occasionally conflicting elements of human needs and site characteristics. The designer's view of a site as an array of understandable, usable systems is the result of knowledge and experience—the traditional academic foundation.

Both client's needs and the users' needs must likewise be understood as systems—political, economic, and personal. The designer's experience with and knowledge of these human systems is equally as important as knowledge of site systems. It has to be obtained from people, the users of the work. Knowledge may be obtained in a case-specific, time-compressed way, through data solicitation and compilation. Limiting the time spent in people contact has some obvious liabilities; the depth of disclosed information may be meager. As mentioned previously, people don't necessarily know themselves very well. Or knowledge may be a deeper, cumulative sort gained by long-term observation and contact with the users. The benefit of this extended contact is obviously the corroboration of things said with things seen. The pity is it cannot be hurried to meet a typical schedule.

The designer's responsibility and most important service is to offer the best professional advice based on complete consideration and synthesis of site and human factors: the fiscal, the physical, the aesthetic, the functional, the political. The designer must foresee the consequences of proposed actions. Decisions may not be the designer's responsibility, but supporting those decisions by appraisal of the outcome is professionally central.

This may not seem like a very definitive set of instructions. It *isn't* definitive—just as standards aren't. Process is our most important product. The most important part of the process is bringing people into it—understanding people's actions, their environmental responses, and their needs, and guiding the physical design process to meet those needs. Later, in the discussion of the site design process, more will be said about blending people's needs (program) with physical resources (site). But consider some checklist items reiterating some previous points to reinforce conclusions:

- Has information been gathered from all ages, sexes, races, and groups represented (*really*—were there people on the street who slipped past categorization)? How about all economic levels?
- Is what was heard consistent with what was observed?
- Is what you have concluded what you *alone* concluded?
- Are trends reasonably represented? Is future potential identified? Future decline?
- Are some trends (fads) *un*reasonably represented? How influential are status symbols?
- What is the relationship between budget capacity and user appetite? Has phasing (and ordering of priorities) been discussed?
- How much has the designer unintentionally influenced the design? Has a direction taken been the result of trying to please? obey? avoid?
- Is there any joy in the apparent conclusions?

The Key Word

Where the limitations and requirements of mechanical devices are determinants, where economic and administrative efficiency is a consideration, where average standards are held out as the magic answer although the action is ostensibly taken to provide for people, the key word is simply *people*. Is the satisfaction of their needs the priority criterion in the development?

Principle 3: Both Functional and Aesthetic Requirements Must Be Satisfied

If we surround ourselves with specialists because we believe in their in-depth knowledge of their subjects, we should expect more from their efforts than quantity. Specialists' products must also have *quality*.

Site design quality can be evaluated on two bases. The first is highest dollar value, that which can be measured in terms of hard cash. Evaluation of highest dollar value is simple. Weigh the relative costs of alternative solutions. If the problem is surface drainage and blacktop will generate the necessary flow, it may be economically foolish to consider brick.

But highest human value—that which is judged in terms of human response—also must be measured. Human value, adding to or detracting from a person's well-being, is found in the intangible influences possessed by every tangible object: in a tantalizing view, the roll of swelling topography, the shade of a tree, or the intrigue of a paving pattern. While the value of human response cannot be price-tagged, it is very real nonetheless.

Matter of Concern

3A. Balance of Dollar and Human Values

To arrive at a quality park design, both dollar- and human-value aspects must be weighed. These aspects boil down to functional considerations (upon which the dollar sign can be placed) and those of aesthetics or beauty (from which pleasurable human response is gained). Therefore, our blacktop-versus-brick dilemma cannot be resolved solely by applying a cost factor.

Function and aesthetics may seem like polar opposites, but they are not irreconcilable. Progress is being made in attitudes toward the value of quality and its payback in human pride and consequent care. Technology continues to shoot forward, providing new materials and applications, and the means (through the computer) of looking at more and more data to test old assumptions and prevailing but unproven opinions. Paper models for "what-if" questions can be built from facts and figures and brought to life in cost-saving practices without compromising the quality of the original design.

Life-cost analysis is one of these forecasting techniques. Carrying cost estimation further than the fracas of the initial bid- and blood-letting, further than the initial cost of construction, life-costing examines the question, so easily lost sight of in the front office, of the cost of the creature throughout its useful life, not just its creation. In go repair, rebuilding, and replacement—many more than three Rs. Some agencies have found first costs may be merely 10 percent of life cost (Figure 2-18 on p. 30).

Requiring appraisal of each design choice in terms of its performance potential imposes a proper responsibility on designer and user to stress unity and simplicity in the planning phase of a project. The advantages of design with these virtues are in the impact of unified appearance on the beholder and in the economic benefits of simplicity of materials—fewer replacements to inventory, fewer processes to train for and track, and greater opportunities to concentrate operations and equipment.

It is important, in striking a balance between dollar and human values, that problems of function and of aesthetics be solved concurrently. Hand in hand. Never apart. (See Figure 2-19.) Aesthetics are never thought of as window dressing applied after function has been solved; function is never treated as an evil forced in after some pretty picture has been established. This "togetherness" in the design process will be illustrated in subsequent chapters, as it will be shown that many moves that can be termed aesthetic actually strengthen the functional side of the solution. And, as design strives to balance dol-

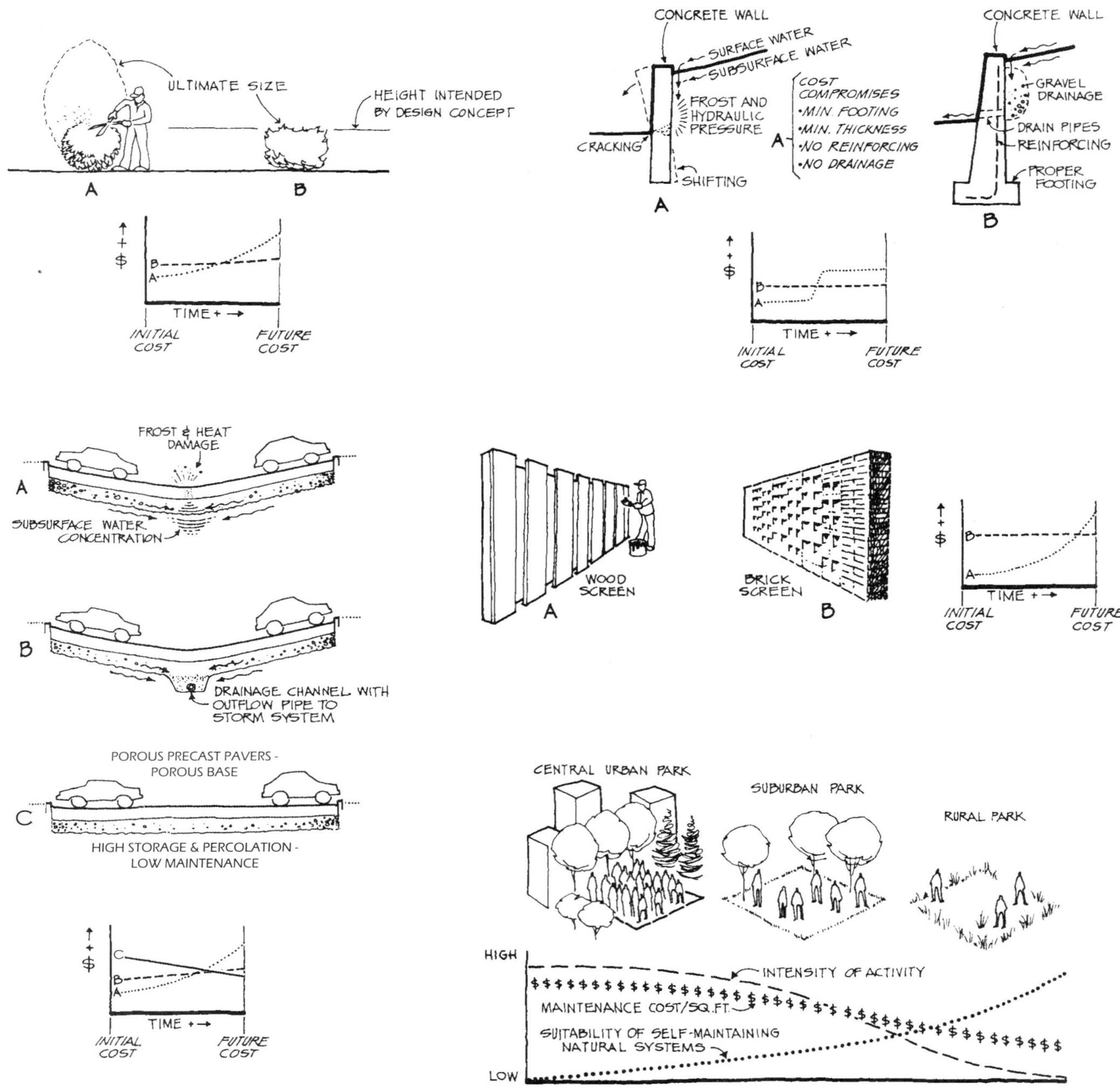

Figure 2-18
Life-cost considerations must reach beyond the cheapest installation. How much will it cost to keep it functional? Try your own charts of some problem areas.

lar and human values, so too should evaluation. The relative success of a scheme should be measured in both contexts.

The Key Words

Every design solution must be workable. That is, every object and system of relationships proposed must function in the most efficient manner possible. Judgments regarding highest dollar value or degree of function can be wrapped around the term *efficiency*. In evaluation, this is the word that provides a test for every one of the site's working parts. At the same time, the word *experience*, in its cerebral sense, can be used to trigger critical thoughts about the aesthetic success of the same parts.

Aesthetic solutions address themselves to the refreshment of the mind as set forth in this definition of beauty: "An emotional response in the mind of the beholder that to him is pleasureful." When this definition is overlaid on Mr. Doell's idea of recreation, "refreshment of the mind or body or both through some means which is in itself pleasureful," the tie between aesthetics and park development becomes obvious, *and* it becomes obvious that the tie is a close one.

Aesthetic quality reaches you through your senses; that is, you must see, smell, hear, touch, or taste something before it can have influence upon you. Accordingly, that which is charged with aesthetic duty must capture the attention of your senses. It must not rely upon contrived intellectualizing for its essence to emerge; it must generate an impact that makes its presence unmistakably felt. On the other hand, if its perceptual message is so weak it remains unnoticed, you will come away from it without gain, the result being akin to the experience of sucking on a straw in an empty glass. Nothing.

Triggered by the word *experience*, measurement of the aesthetic success of any development begins by asking the following questions in turn: Is a sensory experience provided? Is it strong and influential? Is it pleasurable? The relationship between aesthetics and experience will be discussed further in the following chapter.

Figure 2-19
The land drains well and steps accommodate the grade change (function), while the paving patterns satisfy visual appetites (aesthetics).

To More Specific Considerations

Attention to purpose, people's needs, functions, and aesthetics are the umbrella considerations that spread themselves over the entire period of design thinking. The key words: *why*, *people*, *efficiency*, and *experience* focus attention on basic issues with which the designer contends and, therefore, comprise the tests which must be passed by every development commitment.

Nestled under the umbrella are more detailed principles that also guide design decisions. In the ensuing chapters, these will be presented under the labels of aesthetics and function, for it is in the course of solving problems of experience and efficiency that actual project purposes, people's needs, and dollar and human requirements are identified.

Aesthetics and function are separated here in chapters 3 and 4 for discussion purposes only. It must be remembered that in the design process they are interdependent. Likewise, we'll touch on the concept of *sustainability* now and look at it further in Chapter 4.

Matter of Concern

Sustainability, sometimes celebrated as **common sense**, needs to be a shadow presence influencing all the considerations and evaluations made in the design of parks. Hippocrates's advice to doctors allegedly says, "first do no harm." Very good advice, no matter how it's translated. It's good for everybody involved in the park's development and life.

Sustainable Park Development—Key Ideas

1. *Listen to the land.* Nearly all land uses in all parks need level ground. Are plan decisions using the easy, sensible locations or will the site be expensively, extensively cut to pieces for a foolish scheme?
2. *Value compared to cost.* Least expensive building materials and nonhardy, inappropriate plant materials may save money in the purchase phase. The later cost of intensive maintenance and continual replacement is the polar opposite of sustainable practice. Evaluating apparently greater first costs may yield the better, longer-lived investment, with benefits of discoveries in recycled materials and innovative systems.
3. *Water management equals free land.* Water runs downhill (inconvenient, but true). Ideally, after development, the amount of surface water running out of the park should be no greater than it was before development. Yet necessary hard-surface development in the park (buildings, play courts, parking lots, roads, walks) does replace absorptive surfaces with potential runoff accelerators. Greater runoff has to be countered with on-site storage, usually in ponds that slow and hold the water. Better solutions are possible with pervious paving providing water storage in the base. Parking and similar paving with underground water storage allows elimination of retention and detention ponds, yielding more available land for recreational use. The results are sustainability, more active use area, and even the possibility of improving on the retention capacity of the original site. More about this in Chapter 4.
4. *Good maintenance is thoughtful maintenance.* All elements in a park need care. Some care is done on a "this is the way we always did it" basis. Sustainable maintenance should be a process of first determining the objective and finally selecting the most efficient means of accomplishing that objective. Good records and regular evaluation secure this goal. The best maintenance solution results from understanding. Design decisions should be made with complete information shared by all the responsible parties—management, maintenance department, program department, and the design team. The result is a park that makes the intentions of the design real, with a maintenance plan that can keep those intentions real as the park lives on. Of greatest concern should be this last question:
5. *Does the decision about to be made make common sense?*

3 The Aesthetic Considerations

To weave aesthetic quality into a development, a designer applies not only principles of artistic composition but also powers of intuition. The possession of the latter is a must. For in such a complex field as aesthetics, there exist so many factors capable of modifying each other that it is impossible to set down, much less follow, rules that never vary. Yet, it is not necessary to possess this mysterious capacity in order to discriminate between the pleasing and the displeasing. Unschooled observers can readily gauge the aesthetic reasonableness of any development if they simply sharpen their powers of awareness.

The initial step in honing awareness is to establish extremes on a mental "excellence scale" as you go about the environment. Although beauty may be hard to describe, you certainly know when you are under its influence; for instance, you would probably not find it too difficult to rate the relative aesthetic merits of Frankenstein's monster and Michelangelo's *David*. It should be just as easy for you to judge a blank-paved, junk-filled playground against one with stimulating layout and imaginative equipment.

To know the best, you should also experience the worst. Thus, never at any stage of shopping for excellence should you feel you have seen all. Remain alert; always be on the lookout for something a notch below what you have previously considered poorest. But also continue to expect that which is above the best you have heretofore seen. Developments with which you are directly concerned can then be appraised against the extremes in this mental catalog.

To establish degrees of excellence (or offensiveness) among those developments that lie between the extremes, provisions for *order* and *variety* should be measured. While "beauty is in the eye of the beholder," order and variety are psychological appetites that most behaviorists agree are common to all beholders. Regardless of its other attributes, an aesthetic composition must possess these if it is to generate a pleasurable response.

Figure 3-1
Too much uniformity results in monotony.

Figure 3-2
Excessive dissimilarity breeds chaos.

Figure 3-3
Order and variety in balance.

Figure 3-4
Invigorating: an expression inherent in nature.

The need to perceive order or logical correctness about what is experienced stems from the human need to understand and find reason and regulation in human works. Under the influence of order, the mind experiences tranquility. All is right with the world. Everything fits into place. One can go about meaningful activities without being disturbed by the surroundings.

The desire to witness variety or contrast—a touch, but only a touch, of disorder, difference, or change from the expected—is rooted in the need to exercise one's senses. Variety provides excitement and stimulation. It is the spice that combats boredom, keeping the faculties alert, thereby honing them for other tasks of daily decision making.

Yet, overabundance of either in the environment is equally distressing. If regulation is too obvious or there is "too much of the same" as in Figure 3-1, the scheme becomes tiresome and monotonous. Conversely, if contrasts ricochet about in unending fashion as in Figure 3-2, the result is unnerving chaos that feeds mental distress.

In a successful design, both order and variety are present in fragile equilibrium. Note in Figure 3-3 that there is just enough order to suggest logic and stability, coupled with the right amount of dissimilarity among the parts to vitalize the situation. Primarily, it is the designer's intuition that points to the appropriate mix; one may state, "Designing is like preparing a fine stew by ear."

To measure provisions for order, the following criteria can be applied. Whether or not the proper amount of variety has been instilled, however, must be left to the critic's personal judgment. To help tune up your ability to make that judgment, comments regarding variety will be interjected wherever opportunities arise.

Principle 4: Establish a Substantial Experience

The first step toward understanding or sensing organization—perceiving order—in a work is placing a label upon it. It is answering the question: "What is it?" This is why you see titles on abstract paintings. They are the artist's concession to the general public.

Since it is impractical to hang labels on a park development, the development itself must have such strong character that it produces an impression capable of being identified. As illustrated in Figure 3-4, such impressions are distinctly etched in nature. Because of their distinctive characteristics, we have come to label natural units as, for example, prairie, desert, lake, or plain, and are immediately able to distinguish one from another.

The same thing is possible among human works, as is pointed out in Figure 3-5. Each constructed unit is capable of evoking an emotional image whose influence could cause one immediately to label it *peaceful* or *exciting* or *awesome* or whatever else might appear to fit. If the radiating image is strong enough to be labeled upon first contact, it will quickly capture the viewer's attention, thereby maximizing the possibility that the development will provide the experience that the label implies. How do constructed works create such substantial expressions?

Matters of Concern

4A. Effects of Lines, Forms, Textures, and Colors

The raw materials of site design are not trees, land, and paving materials, but things whose presence is as real as those material objects (see Figure 3-6): *lines* (single edges indicating directional movement), *forms* (external appearances of objects defined by lines making closed circuits), *textures* (distribution of lights and darks over surfaces caused by inconsistencies in illumination), and *colors* (qualities of reflected light refracted by the eye's prism). These are the raw materials of any artist, as was inadvertently admitted by James McNeill Whistler when he insisted that we call the famous painting of his mother "Arrangement in Gray and Black." Whether found in Whistler's work or in the trees, land, and paving materials with which a landscape architect works, lines, forms, textures, and colors possess the potential for producing emotional effects.

Consider first the potential inherent in lines and forms. (Since a series of forms rhythmically leading the eye to and fro attain the directional tendencies of line, line and form can be investigated together.) Straight lines are bold and domineering; they move the eye forcefully (Figure 3-7). On the other hand, horizontal forms are peaceful, calm, and restful, for they lie comfortably on the ground at harmony with gravity (Figure 3-8 on p. 36).

Ninety-degree verticals possess a dynamic quality as they move the eye upward; the more attenuated the form, the more forceful the movement, and hence the greater the uplifting sensation of soaring (Figure 3-9 on p. 36). Diagonal and zigzagging lines are active and spirited, for there is lots of erratic movement in many directions (Figure 3-10 on p. 36). Curved and undulating lines are not as dynamic as zigzags. Being slow and meandering, they are inclined to be gentle and tranquil (Figure 3-11 on p. 36). But if the curve changes direction rapidly, it can produce an animated or gracefully spirited feeling (Figure 3-12 on p. 37).

Pronounced rough textures are bold and domineering like straight lines and, in the extreme, ponderous and primitive (Figure 3-13 on p. 37). By contrast, fine textures are inclined to be sprightly and a bit fussy. Since fines are subtler than rough textures, they can produce a more casual effect (Figure 3-14 on p. 37).

Bright and high-value colors are gay, lively, and spirited (Figure 3-15 on p. 37). Deep hues are somber and mellow (Figure 3-16 on p. 37). Neutrals (grays and browns) recede to the background and are therefore useful in separating clashing colors, toning down the potentially hectic effect of many unrelated hues. This is why stage designers often use neutral backdrops against which to play the bright costumes of many dancers. When applied to buildings and other dominant facades, this premise might be equally successful in separating out the conflicting hues of city neon.

Color any development cautiously. Since color appeals to the most primary instincts—consider the attraction of both children and savages to red cloth and bright beads—its sensational qualities are the first influences felt. Thus, where the reading of lines, forms, and textures is essential to the scheme, color must be used with some restraint, for if strewn about with heavy-handed abandon, it will overpower everything else at the scene.

Every material object, whether it be fence, shrub, trash can, piece of playground equipment, or body of water, has line, form, texture, and color. That is, through these elements, all ordinary things have

Figure 3-5
Exhilarating: an expression within the capabilities of a human being to produce.

Figure 3-6
The raw materials of design.

Figure 3-7
Straight lines.

Figure 3-8
Horizontal forms.

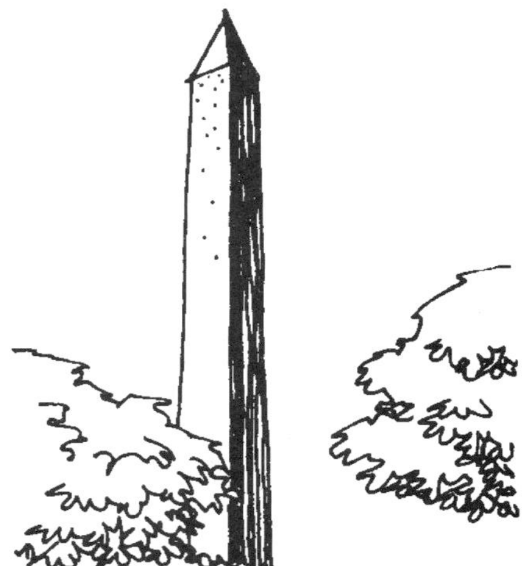

Figure 3-9
Vertical lines.

Figure 3-10
Zigzagging lines.

Figure 3-11
Gently curving lines.

aesthetic potential. It remains for the landscape architect to harness that potential in order to produce an emotional effect.

4B. Effects of Dominance

To exploit their potential, the designer must act from an understanding that the aesthetic elements are never experienced alone, but in relation to one another. It is seldom that the pink bloom of a tree is felt as disturbingly garish, but it may be if it clashes with nearby blues and yellows. Accordingly, the designer must contend with the qualities of all objects seen from every viewpoint in the park.

One reason why you gain only marginal impressions from many developments is that, as you move through them, you do not contact enough objects that exhibit like qualities. For instance, in an outdoor restaurant, there may be a scattering of signboards with bright cheerful colors, but their impact might be watered down by the presence of an equal proportion of dumpy furnishings and somber-appearing material. Having equal visual importance, each quality neutralizes the other, and there ensues little emotional reaction to either.

But what happens when, through creative action, most objects within visual range are made to exhibit similar characteristics? Say, bright cheerful colors on the signboards and sun umbrellas, fine textures in the plantings, sprightly dappled areas of light and dark provided by grillwork, and pulsating water in the fountains. They all add up to an exhilarating atmosphere. Although contrasting qualities might be interspersed for variety's sake, they should occur in limited amounts and remain subservient to the one emotional feeling which has been allowed to *dominate*. As demonstrated in Figure 3-17 (on p. 38), the development now has a coherent expression. It says essentially one thing. It can be labeled. And most important, because the lines, forms, textures, and colors are, each in their own way, making a similar statement, the development's expression stands a good chance of being felt.

However, because visual focus is never static—the eye continually swings about, bringing new objects into view—additional moves to intensify the dominant effect are necessary in order to ensure that it does indeed capture the mind. Measures must be taken to see that the eye focuses only on objects that have the dominant qualities and does not stray to things that radiate other expressions.

In fact, of all the aesthetic criteria, dominance is and has historically been most important. If people, when questioned about where their eye focuses in a designed complex, name a number of specific highlights, dominance has not been well considered. The unity of a place, its vital individuality, is traded for an exciting bench, a clever light, an impressively sculpted shrub, an intriguingly complex paving pattern—all *things* and all *separate*. The more things achieve individual notice, the less compelling the dominant effect and the weaker the personality of the place. The things are remembered. The place is forgotten.

As an example, access ramps appended to an ununified design may be lauded as a satisfactory solution because they are new and suddenly are more obvious than the other pieces in the collection. The result will be embarrassingly abundant attention. If, however, the design is the product of a clear conception—is an entity in which all parts are completely necessary, showing a dominance of unifying forms, materials, and colors—then a ramp incorporated with the same sensitivity that marked the original design will seem to have always been there as soon as it is complete. Its form, material, and color will immediately seem right. Simply, it won't stick out if it isn't stuck on (Figure 3-18 on p. 38).

4C. Effects of Enclosure

The simplest aesthetic move is to wall out the irrelevancies. Consider, for example, the room in which you are reading this material. The walls screen out hallway distractions. You are, therefore, forced to address yourself only to what the room contains, receiving an impression that it is dreary, exciting, peaceful, or whatever.

Enclosure by walls, ceilings, and floors aids in putting the dominant effect across. It does much more. Beyond serving to assist the performance of lines, forms, textures, and colors, enclosure itself has a psychological influence on the confined person.

The effects of enclosure are hardly ever consciously appraised, but you are affected by being enclosed even though you may be absorbed in other pursuits. A parallel can be drawn by an allusion to music. It is not necessary to listen consciously to a tune in order to be caught up in it. Many supermarket managers know this well. Throughout the day, they pipe in dreamy melodies to encourage a slow shopping pace, thereby generating shelf browsing. But near closing time, they step up the tempo with such tunes as "Stars and Stripes Forever," because the help wants you out so they can go home.

Two basic aspects of enclosure play upon the subconscious of the confined. The first is *volume*, or the amount of "emptiness" which surrounds you. Picture the feeling that evolves from and between extremes. Feel yourself encased in plaster. Proceed to being shut up in a closet. Then move to the den of a "typical" home. The degree of comfort generated at these various stages is directly related to the surrounding volume. Arrive at a volume minimally comfortable for a single person, and you have a room suited to introspection: the den.

For contrast, associate yourself with maximum volume: an empty room twice the size of the Houston Astrodome. Consider sitting there, and it should be immediately apparent that this is not a volume suited to meditation. You may have an initial sensation of awe. But this will soon degenerate into discomfort as there arrives the sudden realization of your smallness in relation to the vast amount of volume that surrounds you.

The second influencing aspect is the *type*, or form, of the enclosure. While there are many variations, most enclosing forms can be placed in three basic categories. There is first of all static or complete enclosure. Usually square or circular, a static volume does not give a sense of motion. It is inactive. It just sits there. As such, it is well suited to functions that require isolation or attention to the center.

On the other hand, linear enclosures are elongated volumes that "move" in a definite direction and are open at both ends. Since the volume moves in a direction, so will the eye of the person contained in it, making it logical for him or her to proceed physically on the insinuated tack as well. Therefore, while a linear volume is not conducive to inward-oriented activities, since its motion might be distracting, it is well suited to movement or circulation: a hallway. At the end of its tunnel-like form, you might find a static enclosure: a room. The completeness and lack of movement of the terminating volume signifies a definite end to the journey.

The third type, free enclosure, is a meandering volume that allows movement of the eye in any number of directions. It is suited

Figure 3-12 Spiritedly curving lines.

Figure 3-13 Rough textures.

Figure 3-14 Fine textures.

Figure 3-15 Bright colors.

Figure 3-16 Dark colors.

Figure 3-17
A strong aesthetic statement is made when all objects within visual range are made to exhibit complementary qualities.

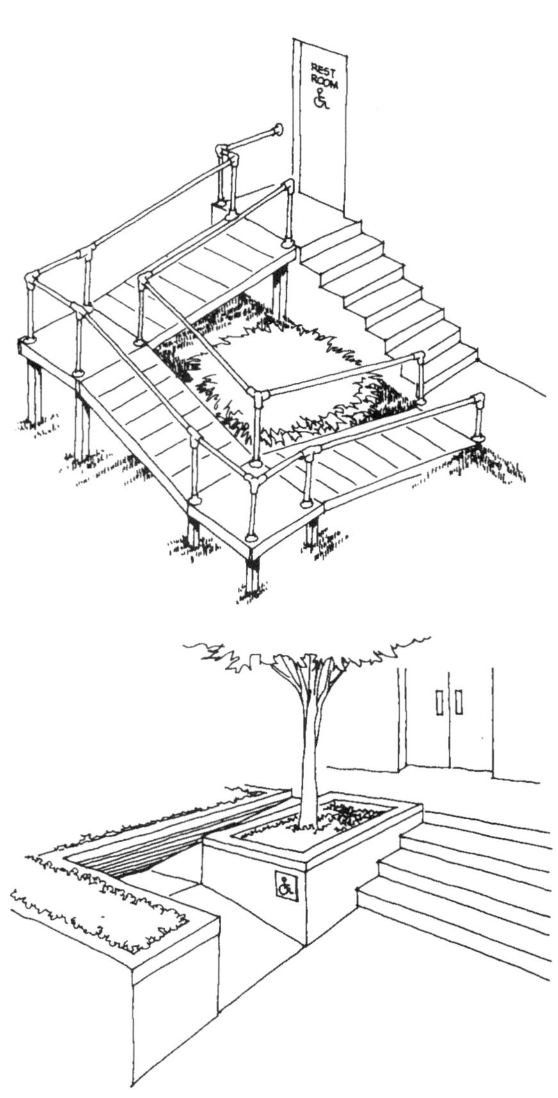

Figure 3-18
Unified design doesn't stick out as if stuck on.

to unregimented activities where individual choice is being encouraged. Free enclosure is seldom found in buildings, although occasionally you may have come across free-flowing "roomless" houses that allow even the children to roam about at will, giving them the impression that all is theirs, too. A house laced with these uninhibited volumes is also a great place to throw parties.

However, the point is not to dwell on architecture, but to use rooms and hallways as introductory examples of what is faced by landscape architects in the outdoors. Such a devious route to the issue appears necessary, for while everyone understands that interior rooms are enclosed, few recognize that the outdoor environment is as three-dimensional as the indoors. (See Figure 3-19.) A prime purpose of outdoor-area design is not simply to devise two-dimensional ground patterns, but to create three-dimensional volumes so that the aesthetic and functional advantages of enclosure can be gained.

As illustrated in Figure 3-20, the base plane takes traffic. Where vertical planes are above five feet in height (eye level), they can be sight, noise, wind, and sun barriers. Below this height, vertical planes still form physical deterrents and thus can help to guide circulation. Overhead planes provide additional shelter from the elements.

Since the planes have both aesthetic and functional ramifications, the two objectives must be addressed concurrently in the design process. Thus, a designer considering the location of a fence must be thinking not only about the size of the volume it will circumscribe and the visual effect to be achieved, but also about its relation to wind direction and about details regarding its construction and maintenance.

The wisdom underlying the concurrent handling of aesthetics and function is further supported by the fact that aesthetically pleasing spaces can serve to make functions more efficient. For instance, static spaces aid in maintaining solitude for passive forms of recreation that would otherwise be destroyed by outside distractions. Static qualities also help the functioning of activities complete unto themselves where inward attention is desired—assisting the concentration of participants, as in lawn bowling or assisting the concentration of spectators, as in an amphitheater.

Spatial types can also give information suggestive of an action. Static spaces provide a logical climax to a journey. The complete nature of the space halts the eye, dashes the urgency to continue forward, and provides the security of knowing that one has arrived. As demonstrated in Figure 3-21, the static space says "Stop." The linear space as shown in Figure 3-22 says "Go" and thus becomes a logical place for a roadway, bridle path, walkway, or similar construction. The free space as illustrated in Figure 3-23 (on p. 40) suggests "Meander." Here the eye has many choices of direction. It is allowed to wander and frolic. Although free space is therefore unsuited to a function where explicit movement direction is required, it serves quite nicely for unregimented play activities where one can do as one pleases, where one is being encouraged to be in charge and chart one's own course.

It is as difficult to think in terms of outdoor volumes as it is to convey their emotional effects if you are not experiencing them at the time. Communication here is also hampered by the fact that the materials that lay up outdoor spaces are far less obvious than those with which a building architect works, and there are many subtle variations that often escape all but the searching eye. For example, in instances where opaque vertical planes are missing, simply lowering a tree canopy creates an intimate *sense* of enclosure. (See Figure 3-24 on p. 40.) While lacking sides, what is created is still for all essential purposes a container. Its form may not be consciously seen, except perhaps by a student of such things, but it is surely felt by those within.

Whether or not you can trace out a form is secondary to the degree to which you can feel the sensation that a space conveys. However, to reinforce the substance of this chapter in your mind, you may wish to seek out both. Accordingly, in any number of outdoor circumstances in which you find yourself, observe what materials make up the overhead, vertical, and base planes. Although comprised of earth and plants, you may find that the spaces are as hard and as defined as architectural spaces and can be readily categorized as purely static, linear, or free. Or they may possess merely approximate form, which, if accompanied by sensation, serve their aesthetic purposes as well as those of more obvious definition.

Appraise the impact of the sensations. Sit under the low-hanging branches of a crab apple tree in an enclosed courtyard (Figure 3-24). Is it comfortable for retreat, meditation, subdued conversation? Then place your chair in the middle of a football field or, even better, the Bonneville Salt Flats if it is handy (Figure 3-25 on p. 40). Feel any difference? Now walk down a road flanked by Lombardy poplars (Figure 3-26 on p. 40). Are you compelled to move forward? Afterwards, stroll into a flattened wheat field (Figure 3-27 on p. 40). Have you now lost the urgency to strike out in a definite direction? Finally, survey a panorama of rolling valleys laced with undulating brows of mature trees. Does a refreshing and uplifted feeling come upon you as your eye meanders through the animated volumes? Move down into the nearest glen. Are the initial sensations retained?

This brings us to another point. Just as the functional areas of the site must be considered one in relation to another, so too must the spatial experiences provided. Nobody is plopped into a space from above like a chess piece. A person moves through a space. Thus, the impact of a space seen prior to coming upon another must be considered, which means that what the designer strives for is a carefully conceived complex, a sequential organization of related views and sensations. Unless spaces are so connected, much of their impact can be lost in the intervening gaps. The fact that they are disconnected is another reason why many outdoor spaces go unnoticed.

Figure 3-19
The outdoor environment is three-dimensional.

Figure 3-20
The planes that enclose outdoor spaces.

Figure 3-21
Static space.

Figure 3-22
Linear space.

Figure 3-23
Free space.

Figure 3-24
An intimate space secured by the low tree canopy.

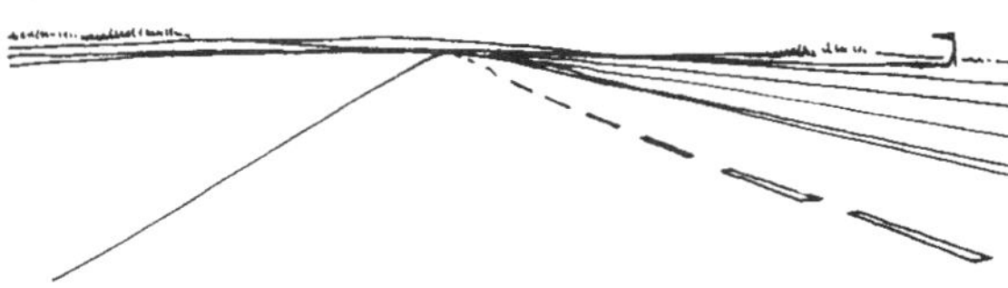

Figure 3-25
The unrestricted overhead plane denies a sense of comfort to the space.

Figure 3-26
Strong vertical planes provide a message of movement to the eye.

Figure 3-27
Lack of a vertical definition creates mental ambivalence.

These awareness tests should make it easier for you to understand why a site development plan cannot be adequately appraised as a two-dimensional pattern. In the critic's eye, the site should leap from the paper as an interlocking sequence of volumes, each space serving a function, each outside "room" conveying an emotional message, for this is the reality of the plan once it is built. (Compare Figure 3-28 with Figure 3-29.)

In the appraisal, however, remember that the full experience will not come from the quality of the spaces alone. As significant as their impact might be, the spatial planes form but the aesthetic skeleton of the site plan. They are only bare walls that remain to be adorned with lines, forms, textures, and colors. It is the grand sum of all the emotion-producing elements that produces the experience: the effects of enveloping space coupled with the characteristics of the lines, forms, textures, and colors of the defining planes and of the objects in the space. Substantial experience comes from this complementary contribution. Without enclosure, the effects of lines, forms, textures, and colors are watered down. But without the aesthetic elements, the spatial compartments will have a sterile look.

Principle 5: Establish an Appropriate Experience

While order may be advanced through a dominant effect, it is not ensured unless the effect can be sensed as appropriate. Stated in another way, it is not enough simply to feel an impact and know what it is. The answer to the question "Why is it?" must also be apparent.

One answer to "Why?" might be to provide an experience that extends something the viewer already understands. Such a provision accommodates order by drawing upon a person's tendency to associate one thing with another that logically belongs with it: hotdogs and beans, Reggie Jackson and baseball, but certainly not ice cream and sauerkraut.

In site development, surmising that what exists is already understood and therefore accepted, the landscape architect strives to transfer the aesthetic qualities of that which is already present to that which is proposed for new construction. As will be discussed, the existing qualities from which the designer draws these clues of association may be something about the physical character of the site, the personal makeup of the user, or the atmosphere commonly associated with the function. If the landscape architect were to ignore such clues, he or she might produce a blatant sore thumb. The beholder then would not only be disinclined to embrace it, but, not understanding why, would be quite likely to hold it up to ridicule.

This in no way rules out the unique. Indeed, it has been stated that variety and new experiences are essential to everyone's well-being. It does suggest, however, that that which is foreign may be more readily accepted, hence more meaningful, if it is rooted in something that is familiar. This is quite in keeping with the point made earlier about dominance. To repeat: "While there might be contrasting qualities interspersed for variety's sake, they should occur in limited amounts so as to remain subservient to the one emotional feeling which has been allowed to dominate." In addition, while this principle is addressed to the provision of order, it will be shown that in its own ambivalent way, its application can also ensure variety.

Matters of Concern

5A. Suited to Personality of Place

Locations have personalities or pervasive moods that can be described as, for example, awesome, exhilarating, or peaceful. Where these dominant effects are created by nature, everything contributes to the experience. Little seems out of place. When people interject their things amidst this order, there is disruption. But the fact that constructed inclusions are concrete, steel, and tailored beams hardly constitutes a license to dismember the prevailing influence. Rather, here is the opportunity, indeed the demand, in the interest of avoiding chaos, to wed human works to existing conditions. This is actually common sense, for it is easier to provide a substantial experience through an intensification of what already exists than to first water down a pervasive mood and then begin all over again with the insertion of a feeling created from scratch.

There are two basic ways in which human inclusions can be made compatible with their place. The first may be called *physical extension*. That is, new features can be positioned so as to appear to grow organically from the site. Structures may either be tucked into existing corners as indicated in Figure 3-30 or be made to rise from crests as illustrated in Figure 3-31 (on p. 42). The former is a rather passive acknowledgment of prevailing mood, whereas, through exaggeration, the latter accents and hence intensifies an existing emphasis. The second method works through an awareness of the *factors that give the place its personality*—our old friends: lines, forms, textures, and colors. As found in existing vegetation, soil, minerals, and other elements of the original landscape, these qualities may be repeated in the construction materials that go into the new one. Or dominant forms such as the horizontal sweep of the prairie or the sharp angles of mountain peaks can be reflected in related human structures.

This concern is as applicable to natural areas as it is to urban situations, for in both cases it boils down to a desire to achieve transition from the existing to the proposed. In the woods or wherever human structures are quantitatively minimal, the thrust is to make human works appear to be a part of their natural surroundings. Thus, where nature predominates as shown in Figure 3-32 (on p. 42), the call is: *Blend people with nature*. Where buildings and pavements are abundant, there remains no less an opportunity to make developments compatible with the old, for lines, forms, textures, and colors of surrounding structures can readily be reflected in the details of the new works. Or the new may be made organic extensions of what exists through the creation of viewing channels to significant features outside the construction limits. Thereby, one is tied to another, making both seem related parts of a whole. Where people are responsible for both the existing and the proposed as illustrated in Figure 3-33 (on p. 42), the parallel call might therefore be: *Blend the new with the old*.

This urge to effect a transition often directs the selection of locations for certain facilities, for it becomes easy to tuck a facility into the site if the proposed location approximates the form of the new structure. Existing hollows are ideal for amphitheaters; valleys are well suited for roadways; broad flats are perfect for playfields. The matching of function to existing site form also minimizes installation costs, for to effect the blending of the new with the old, all that is required is minor reshaping. (See Figure 3-34 on p. 43.)

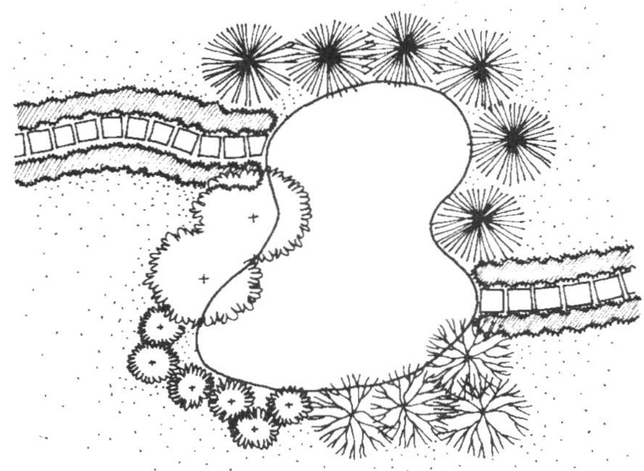

Figure 3-28
A two-dimensional reading of a site plan is misleading.

Figure 3-29
The plan must be interpreted in terms of its three-dimensional reality.

Figure 3-30
The constructed element is made to grow out of its surroundings.

Figure 3-31
The constructed element is sited so as to extend the drama of the existing topography.

Figure 3-32
Nature's textures are repeated in the material of the human structure.

Figure 3-33
The patterns of the old building are reflected in the new.

5B. Suited to Personality of User

In the transition from old to new, creativity can provide certain twists to ensure that, at the same time, effects that reflect the personality of the user are interjected. This might be called for in private places where individual traits are easy to discern. For an individual of conservative bent, only clean, sharp, no-nonsense design devoid of excessive ornamentation might be desirable, since it is likely that this individual is most comfortable in this type of environment. Yet the severity of such design would probably unnerve a more ebullient personality. Such personalizing is well in keeping with the desire to surround a user with the familiar, for what is dearer to a person than one's own self? Success in this can foster a most intense sense of identity through which pride may be generated.

The premise can be equally well applied to public places (Figure 3-35) that are used by homogenous groups, such as ethnic groups or those with strong regional traits. Flamboyant design seems reasonable for the public squares of the extroverted, and the daring and unorthodox are forms that the pioneer-minded might happily embrace.

Personalizing areas with characteristics familiar and appropriate to the users creates a powerful base of identification with and pride in a development. This in turn can generate the caring attention essential to good maintenance of the area and respect for it. The landscape architect has to strive for this personalized quality yet not lose sight of function and efficiency.

5C. Suited to Personality of Function

If we stretch definitions a bit, we will find that activities also have personalities that can be translated into physical development through design. Recognition of such a thesis is useful in situations where the personalities of users cannot be readily determined, for it can be assumed that users will go to a place in a frame of mind associated with the anticipated activity. It follows that meditation will be enhanced if it takes place in a peaceful environment (Figure 3-36), perhaps a static space with cool colors, fine textures, and placid waters, while playground action will be heightened in a free space with bright colors and other features that add up to spirited and active surroundings. This placing of function in an appropriate setting intensifies user enjoyment. By aesthetic suggestion, one is encouraged to engage in the activity more fully.

At the same time, however, recall the importance of mental exercise. The designer can enrich projects with unexpected variations, fascinating details, or exciting surprises: an unanticipated opening to a view, a gurgling pool appearing in an apparently tranquil setting, or a suddenly discovered richness of material in a hidden place—all potentially stimulating mental experiences because of the unexpected contrast they provide. These small touches of contrast are sometimes the toughest things to address in a design because to balance them requires great sensitivity. If they become too frequent, they slip from contrast to confusion, killing the dominant effect. Yet they are important. Without some contrast built into the design's personality, that personality may be so functionally correct that it will be unforgivably dull.

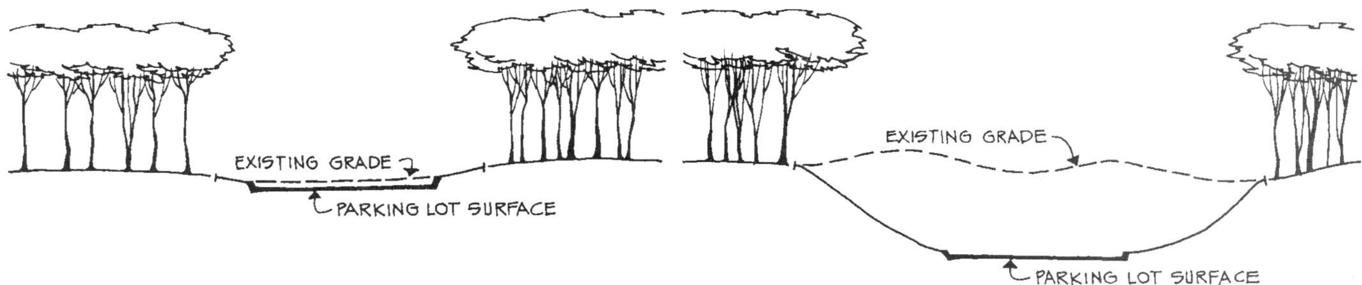

Figure 3-34
New construction can be readily fitted to the form of the land if the original surface approximates the desired finished grade (left), but extensive disturbance occurs where there is a major disparity between the two surfaces (right).

5D. Suited to Scale

Scale is a relative unit of measurement. As expressed in physical features and spaces, its appropriateness must also be sensed in a quality work. Designers of outdoor facilities concern themselves with two types of scale.

The first is *human scale*. To understand our world, people have a tendency to measure things in terms of that with which they are most familiar. The most basic "known," hence the most widely used unit of measurement, is one's own physical self. If one is to be comfortable, there must be things in a space that can be mentally measured in terms of one's own height, arm length, width of finger. If these are not provided as shown in Figure 3-37 (on p. 44), people tend to become confused and cannot comprehend. They are overwhelmed. Set yourself in the midst of the Grand Canyon and you might be awed, but you will remain uneasy, for your sense of security will be affected. You will not be able to stand it for any duration. To stem this anguish and gain mental comfort, you need to be surrounded by things of human scale, elements that can be measured in relation to your own self. (See Figure 3-38 on p. 44.) This holds true for any place where a person is expected to linger.

Speed scale is also of design concern, for the swiftness with which you move affects your ability to experience. If you are going at 60 miles per hour in your car, you can only distinguish the large shapes, sharply contrasting textures, and great masses of color illustrated in Figure 3-39 (on p. 44). Details cannot be comprehended.

If you are walking, however, the slower pace enables you to experience the greater intricacy shown in Figure 3-40 (on p. 44). And when you stop, you are able to investigate and understand all textural and color subtleties. In fact, when you sit, you may subconsciously demand such nuances as indicated in Figure 3-41 (on p. 44), for in their absence you may become bored. Here is another example of aesthetic direction shaped by functional realities. The function, in this case the type of movement, supplies the purpose behind the aesthetic decision.

Figure 3-35
A public place.

Figure 3-36
A place well suited for meditation.

Figure 3-37
Discomfort: things of human scale are missing.

Figure 3-38
Security: human scale is present.

Figure 3-39
While one is moving at high speed, only masses can be experienced.

Figure 3-40
While one is moving at a walking pace, details can be comprehended.

Figure 3-41
When one is sitting, detail intricacy may be demanded.

Understandings and Habits

The tying of development to place, user, and function and the consideration of scale make design visibly logical, for the experience provided is sensed as appropriate. It fits. The preceding can also be used to illustrate other points in retrospect: with attention to these matters, design gains *purpose*; it is addressed to satisfying the senses of *people*; it interweaves *aesthetics* and *function*, for aesthetic direction is supplied by insights related to how the site will be used. In other words, all the tests posed by the umbrella principles are passed.

In addition, while instilling order, moves to ensure the appropriateness of experiences—especially those moves that seek to preserve the existing character of a site—serve variety as well. Just as people vary, so do pieces of land. Like a person's fingerprints, each site is unique. Therefore, measures taken to see that the new extends and intensifies present site qualities are steps that will guarantee that the site *remains* unique.

Too many people are blind to this simple device for maintaining what is left of environmental diversity. Those who dump hilltops into valleys or, even worse, obliterate forests with Coney Islands deserve no better than this upon their headstones: "What you have made may be found in many places; but what you have destroyed is found nowhere else in the world."

What might also be carried over from this chapter is awareness of some essential creative purposes. It is the designer's task to place functional areas in appropriate spaces on the site. If such enclosures are only vaguely defined, the landscape architect moves to reinforce their structure so that their experience-producing potential may be realized. Where outdoor rooms of desired type or form do not exist at all, the architect's charge is to create from scratch spaces to serve the functions. Once the spatial skeleton has been established, the task is to finish off the experience by providing a composition of lines, forms, textures, and colors appropriate for each spatial compartment.

In addition, as you appraise a designer's work, it will be useful to understand the implications of the aesthetic principles. Ask such questions as: Is an experience sensed immediately; does it come on as a definite impression? Is there variety; does the experience remain fresh after many journeys to the site; are new perspectives continually discovered? Give it the second- or third-journey test to ensure that intrigue remains after novelty has worn off.

Is experience appropriate to the function it serves? Or is it completely out of whack with what is going on: a honky-tonk frivolousness in the midst of passive pursuits or placid surroundings for hustle-and-bustle activities? Is there adequate transition between old and new? Is development sensed as an organic extension of its surroundings, or is it a comical sore thumb?

To sharpen your critical abilities so that you can spy the clues that will lead you to ask these and similarly penetrating questions, form some general habits. Compare experiences on your excellence scale. Are they perhaps akin to Figures 3-42 and 3-43? Negative or weak experience means that the designer has not done a good job, whereas positive experience is evidence of the landscape architect's ability to provide refreshment of the mind.

Look at things in aesthetic terms or in terms of the qualities that add up to experience. See that shrubs, just as much as fences, pavements, or earth, have texture. Note that a mass of trees can form defining vertical planes just as readily as walls can. Decide for yourself about the sensory potential of three-dimensional volumes in the outdoors. (See Figure 3-44, next page.)

Search out these qualities. The development of awareness and critical sensitivity is what this chapter is all about, along with emphasis on the fact that aesthetics have purpose. These judgments are not frosting-on-the-cake afterthoughts. Ask: Why this color? Why this texture? Why that particular three-dimensional configuration? Why that opening in the space? Why such complete enclosure? Expect clear, logical, and convincing answers.

Figure 3-42
Rate this scene on your excellence gauge (and, per Duke Ellington, "It don't mean a thing if it ain't got that swing").

Figure 3-43
Where would you place this vignette on your experience scale?

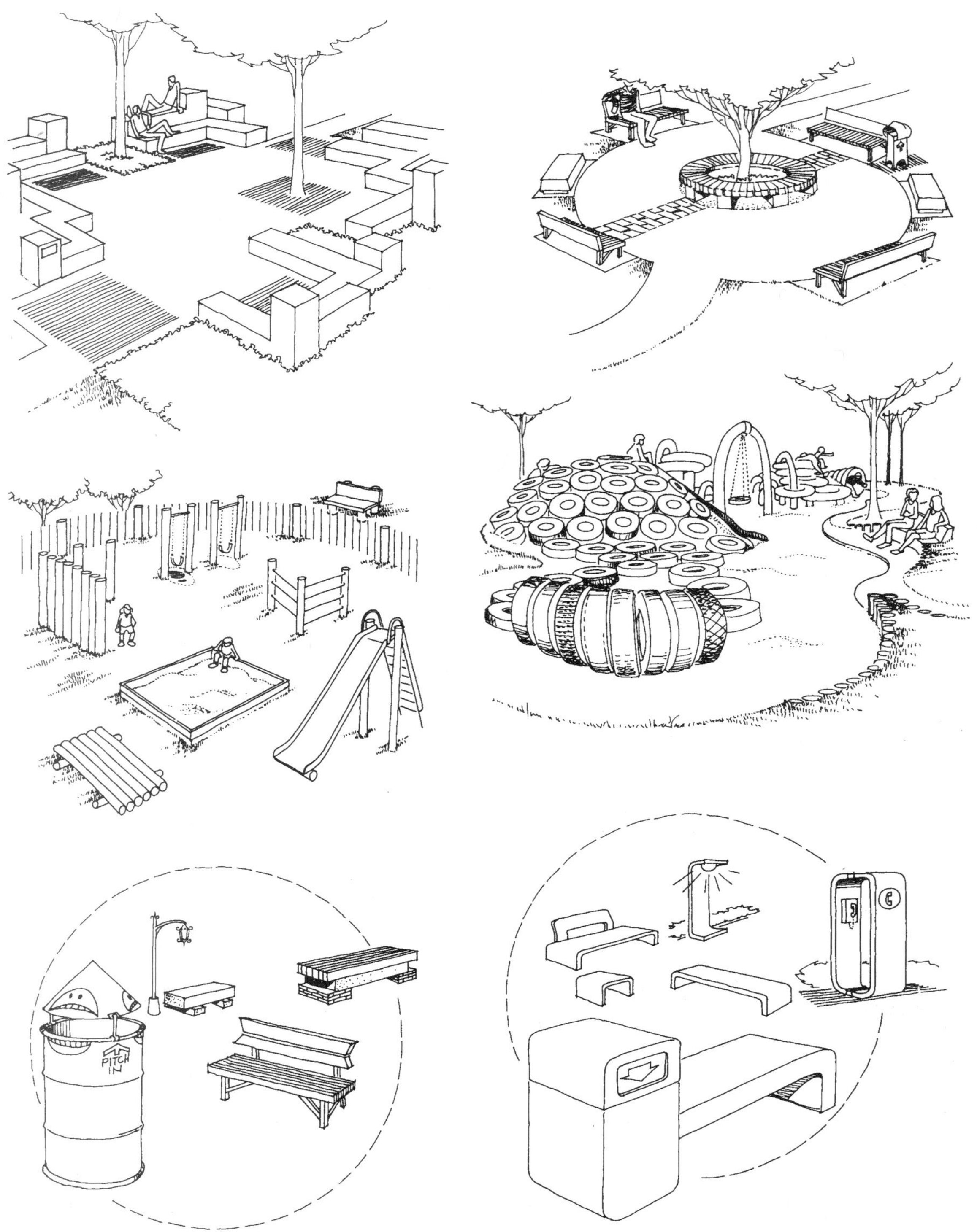

Figure 3-44

Some more examples for you to rate. Look at each of the three sets of schemes above; compare left to right. Seek out examples of criteria offered in this chapter. Pick the winner in each set.

[Top Set] Seating Mini-Park

Left: Consistent forms and materials; space has clear entrances and exits, with unobstructed way through, yet aims the visitor at the seating pockets; seating provides much variety within the consistent forms. What else?

Right: Inconsistent variety of materials and "stuff"; apparent route in-out runs smack into tree-bench combination. Comfortable seating at the tree? What else?

[Middle Set] Children's Play Area

Left: Scattered, unrelated play equipment. Parent bench outside the fence. What else?

Right: Consistent forms; good play potential; parent seating integrated into total form, good observation opportunity. What else?

[Bottom Set] Furniture Selection for a City Park

Left: No relationship between pieces. Some ornate, some woodsy, some "garbage-barrel rococo." Assorted materials difficult to maintain. What else?

Right: Consistency in forms and details; maintenance compatibility in surface care and tool requirements. What else?

4 The Functional Considerations

Unlike Whistler's art, landscape architecture must be used as well as experienced. Its beauty is irrelevant if it is not functional. Functional efficiency can be judged almost entirely on the basis of tangible evidence, for in contrast to aesthetics, where much analysis is qualitative and depends upon sensing the feel of the thing, the functional effect of design is predictable. You can focus on an issue of workability—say, the movement of traffic from point A to point B—put yourself in the place of the user, maintenance worker, administrator, or whoever else is going to be affected, and go through all the possible motions of use to determine the functional reasonableness of the scheme.

What is best or highest quality is a product that has no weaknesses. While there should be no letup in pursuing this goal, it may not always be possible to achieve it. Weaknesses can stem from the unavoidable inability to satisfy conflicting requirements: a demand for a certain provision, but a budget unable to stand the expense; the need for a football field, but a site with precipitous topography. Compromises may have to be found; the critic must be able to distinguish these from questionable conclusions that have resulted from either mental laziness or lack of problem-solving ability on the part of the designer.

The knack that can enable you so to distinguish springs from a continually growing knowledge of what is possible. Thus, it is suggested that you constantly compare park developments and the workability of their parts with a view to developing a *functional* excellence scale to go along with the aesthetic measuring device already discussed. Extreme cases will be easy to ferret out: It works, or it cannot be used at all; it's the right size, or it is obviously too small for its purposes.

As useful tools in determining degrees of quality, the following issues deserve concentration. While each job will have its unique problems, these are the issues generally faced in all instances. As common concerns to which solutions must be found, these are the major points a critic must chase down to conclusion in order to determine the functional quality of a plan.

Before going further, try this experiment using "Google Earth" (and if you don't have it or have access to it, it's time you should correct that situation). In Google Earth, locate your favorite park, a place you know well. Zoom in to a view of the whole park. Looking at the aerial view (the live plan of the place), try walking the paths, driving the roads or lanes, sitting in the spaces, exploring the whole place. Think about what seems satisfying, successful, enjoyable. Think also about elements that are the reverse, negative of satisfying, successful, enjoyable. Try to do it as though you're the maintenance chief, the security director, a parent with kids, or any of the other persons using or keeping the park. Remember your impressions and ideas.

After you've completed this chapter, go back and do the exercise again. Your list will probably be longer and it will hopefully reveal some significant surprises.

Principle 6: Satisfy Technical Requirements

Let's start with the most elementary matters faced in design—those easiest to observe on the site or on drawing paper. These are the minimal standards of quantity, structure, and performance that must be met if the product is to be at all usable.

Matters of Concern

6A. Sizes

We have said that it makes sense to locate facilities on portions of the site where only slight remodeling of the topography will be necessary. This, along with the self-evident need to ensure adequate elbow room for all the functions, means that the landscape architect must test proposed locations against use-area sizes before he or she can proceed to satisfy more complex requirements. Size information comes from many sources, including manuals developed by park agencies, university extension services, and recreation researchers. The sampling shown in Appendices 1, 2, and 3 should give you an idea of the kinds of data which are available, while more extensive lists are found in many of the references cited in the bibliography.

Size recommendations for playing fields and court-game areas can be accepted with little question, for such measurements are determined by the rules of the game. However, size suggestions for many other use areas such as "tot lots" and picnic areas can be modified as design purpose might require, for they are based on rather fragile precedents.

For instance, a widely used standard indicates that a tot lot should have 2400 to 5000 square feet of surface. This standard has been determined by averaging out a sampling of lots in several cities across the country. As reasonable as it might prove to be, its adequacy should always be checked against the type of equipment, mode of circulation, need for buffer, and other demands of the specific lot to be addressed before it is given a blanket acceptance.

Individual design requirements, then, become an understandable basis for deviating from a standard size suggestion. In addition, as will be demonstrated, existing *site characteristics*, *design moves* that may be employed, and *agency policies* can be equally valid reasons for departure. Consider the standard that suggests ten to fifteen picnic sites per acre (4356 to 3904 square feet each). If each picnic area is to be placed upon a base that is stabilized with blacktop or gravel or has privacy screening, more areas may be in order. However, if the site has erosion-prone soil, canopy trees that will suffer under concentrated foot traffic, or steep slopes susceptible to gullying, fewer picnic areas, if indeed any, would be advisable.

Several dual-use concepts can make areas smaller than standards suggest adequate for many uses. One is the school-park policy whereby play facilities used by schoolchildren during the day are released to the neighboring public in the afternoons, evenings, weekends, and summers. Since the same area is used by both school and park patrons, facility duplication is minimized. Thus, 40 acres of dual use, the school-park, does the same job as 60 acres divided between a school playground and a park. (Design organization that allows a school-park concept to succeed is illustrated in the chapter on plan evaluation.) Agencies can also put the same patch of ground to several uses where seasons of play differ; tennis courts can be flooded for winter skating, and football fields can revert to softball diamonds in the spring, to give just a couple of examples.

Standards are therefore viewed by the designer as guides and points of departure. Whereas some might be relatively inviolate (a tennis court requires 6000 square feet), others may be modified in accordance with an analysis of the situation at hand. In addition, many use units, such as passive recreation retreats, cannot be standardized. For these, the amount of site best suited to the purpose usually dictates its eventual size. Caution: In considering deviations, understand that arbitrary adjustments can lead to poor quality. Have you ever seen campers jammed elbow to elbow in violation of the standard that proposes four to seven units (10,890–6223 square feet each) per acre? As seen in Figure 4-1, the attractiveness of the area that drew the camper in the first place has been wrecked.

Figure 4-1
Inadequate space between campsites subverts the camping experience.

In order to determine size adequacy, you might first check proposals against standards. Where manuals are unclear or where deviations exist, the application of the key word *why* becomes useful. It should precipitate an explanation of the designer's rationale, which you may accept or reject.

6B. Quantities

Each park contains a number of things—use units and physical objects within each unit. Judgments regarding the appropriate type and number of units usually evolve from the demand studies mentioned when we discussed the umbrella considerations. Thus, if five tennis courts or one hundred boat slips are called for, it is because it has been determined that the demand for tennis or boating will cause that many to be used constantly. In most cases, such judgments are only best guesses and have to be reevaluated once actual participation is measured after installation. This is why it is always good practice to provide for expansion of those facilities whose popularity might be expected to increase in the foreseeable future. It is important to recognize that, since the turn of the century, the demographics of the population determining the desirability of different uses in parks is changing toward a larger proportion of upper age levels. The change is very important. In 1980, the percentage of the US population over 65 was 25%. In 2000, it was 35%. By 2020, it's projected to be 55%. This becomes a hugely significant number of people at a more advanced age with possibly more time to play or simply exercise, and more need

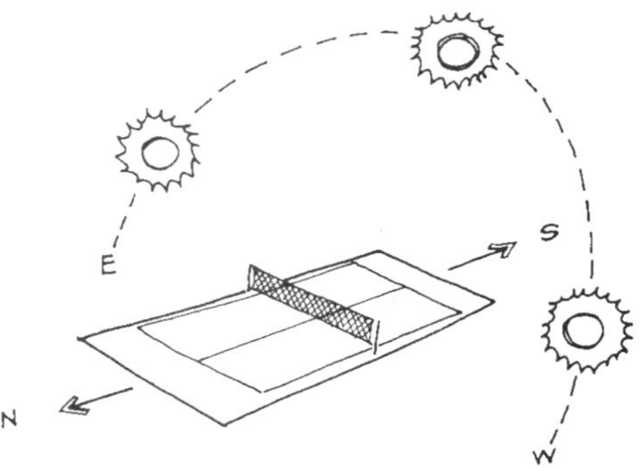

Figure 4-2
Orient tennis courts perpendicular to the sun's course.

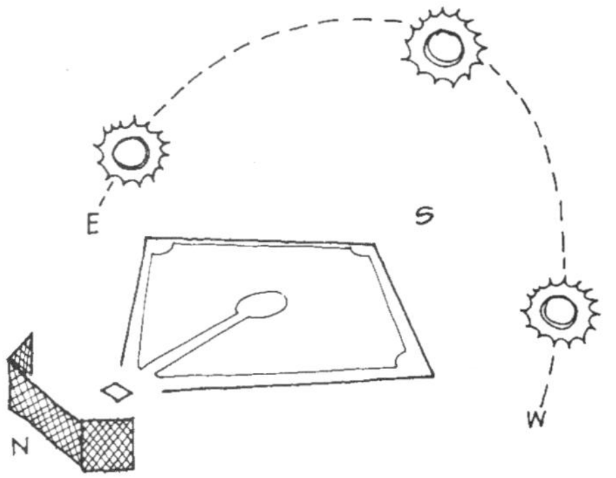

Figure 4-3
Lay out baseball diamonds so that the sun is not in the eyes of the batter.

Figure 4-4
The sun should be at the viewers' backs.

Figure 4-5
Beaches should receive full solar exposure.

Figure 4-6
Eastern slopes are ideal for camping where morning dew and afternoon heat are undesirable factors.

for more places to use for these activities. The evident result is a need for flexible spaces to accommodate multiple uses and multiple ages in overlapping, compatible settings.

The number of physical elements considered adequate for each park unit is usually obvious, such as two posts, a net, and nearness to a drinking fountain for a public tennis court. A few others have been identified for us in the ubiquitous standards. (See the appendices.) As we have pointed out, many standards can be accepted with little question: A family-sized picnic area might very well contain one table, a fireplace, and a trash basket shared by adjacent picnickers; a campground should reasonably have one toilet for every thirty campers. On the other hand, as discussed in chapter 2, the suggestion that the preschool play area should have only the standard chair swings, sandbox, small slide, and simple climbing device may be construed as lazy design.

In evaluating the adequacy of quantities, therefore, attention should be focused on the count of activity units in relation to anticipated use and the number of physical elements necessary to facilitate that use. But functional quality does not evolve merely from providing adequate room and the proper number of things, as the perpetrators of so many "asphalt jungle" playgrounds and other sterile developments would like us to believe. Even under a self-evident principle such as meeting technical requirements, there are additional matters to square away.

6C. Orientation to Natural Forces

First of all, consider the effects of sun upon activity. To keep blinding rays out of participants' eyes, tennis courts and other play areas where a ball is sent back and forth should have their playing axes laid out at a 90-degree angle to the sun's daily course, or essentially on a north-south line (see Figure 4-2). For ball diamonds and other facilities where the missile's course cannot be predicted, priorities must be established since all players cannot be given equal protection. Taking baseball for example, a line from the plate through the mound to second base is the axis usually oriented perpendicular to the late afternoon sun (when the sun is lowest, hence its rays most hazardous) since this protects the batter, catcher, and pitcher (see Figure 4-3). A similar premise applies where spectators are of primary concern. As shown in Figure 4-4, a viewing station should be oriented so that the sun remains at visitors' backs during peak hours.

The sun can also be a useful force, and swimming beaches and garden plots should be oriented to either south or west in order to be given the advantage of maximum solar exposure. (See Figure 4-5.) An eastern orientation is ideal for most campgrounds, for as illustrated in Figure 4-6, the morning sun can quickly erase overnight dampness, yet its heat is replaced by cooling shade in the afternoon.

Wind is another natural force that can affect activity efficiency, as is only too obvious to the players who must lean into the gales and chase the frolicking fly balls in San Francisco's Candlestick Park. Ballparks, tennis courts, and other units where missiles are caused to fly about do not belong in the direct course of intensive wind currents. As demonstrated in Figure 4-7, the direction of wave-billowing winds should also determine the location of boat docks. However, wind blockage or lack of atmospheric motion can decrease camp and picnic ground usability, causing a pall of cooking smoke to hang continuously in the air as if to remind us all of Los Angeles, the ultimate example of inattention to the effects of natural forces in the location of human works. Therefore, as illustrated in Figure 4-8, every opportunity should be taken to orient such grounds to available breezes; this criterion should also be used to determine the placement of other activity areas in humid climates.

Park structures for permanent human habitation such as offices, concessions, rinks, gyms, and meeting spaces require heavy use of heat- or cold-producing energy to resist sun, wind, and winter cold. The site selected and developed for such structures can offer protection if orientation and landform are considered at the early planning stage. The proper combination of earth sheltering and solar design can open the building's interior to warming south sun in winter yet insulate against punishing summer heat with a cool earth jacket around the bottom, sides, and top of the structure (Figure 4-9). Even existing buildings, structurally unable to support soil piled on or over them, can be made at least less vulnerable to energy loss by creative use of plantings for wind buffering in winter and shade protection in summer (Figure 4-10).

The amount of annual rainfall and the periods of drought and downpour present clues as to the seasonal availability of water for wells, fluctuation in the level of water bodies, and flooding possibilities in the lowlands. The absence of underground reservoirs and presence of stream-bank instability and bottomland inundation can be decisive considerations in locating many facilities.

So, too, can conditions caused by snow, especially when coupled with other matters of orientation. For instance, because of its relation to wind and sunlight, the lee side of a hill and its north-facing slope are where snow loads will be greatest and remain for the longest period of time. Figure 4-11 therefore suggests the most appropriate location for ski and toboggan runs in temperate zones. It also indicates where roads that must be kept open in the winter should not be placed.

6D. Operating Needs

Appropriately balanced by attention to personal needs, the requirements of cars, boats, and maintenance equipment must claim their share of the planning. The fact that the minimum turning radius of a car is 20 feet will dictate facets of road and parking lot layout. The knowledge that the maximum grade for a launching ramp is 15 percent indicates what has to be done to the marina bank. The understanding that a power mower cannot negotiate slopes over 33 percent leads to decisions regarding the limits of grassy pitches. Information like that illustrated in Figures 4-12, 4-13, and 4-14 (all on p. 52) is gleaned from technical manuals and, where these are not available, from measurements taken while the machine is in operation.

As a focus for design concern, *operating needs* can also refer to administrative procedures considered essential to the efficiency of the recreation complex. This is especially germane to the operation of complicated special-use areas such as marinas and zoos, where each administrator has strong ideas about how the complex should be run. If the zoo manager considers it most efficient to do most maintenance during those hours when the public is not allowed on the grounds, the designer can opt for a service circulation system that makes generous use of public walkways. However, a network of drives apart from pedestrian ways might be more advisable if heavy service is required when the public is present.

Most operational decisions are determined through personal interviews between the designer and those who will be running the

Figure 4-7
Boat docks should be located out of the path of water-churning wind.

Figure 4-8
Picnic areas require breezes to take away cooking smoke.

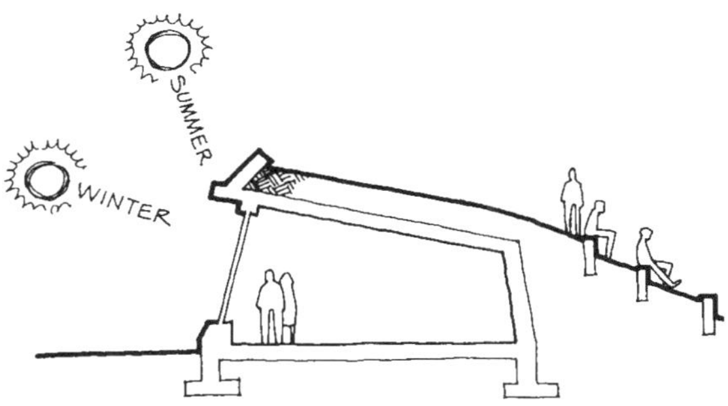

Figure 4-9
Earth-sheltered structures offer natural insulation plus usable surface space.

Figure 4-10
Buffering structures from winter winds and trapping winter sun can reduce energy loss.

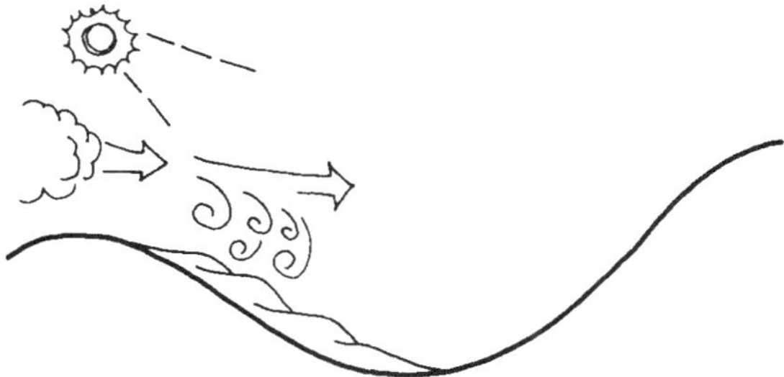

Figure 4-11
Heaviest snow loads occur on the lee side of north-facing slopes.

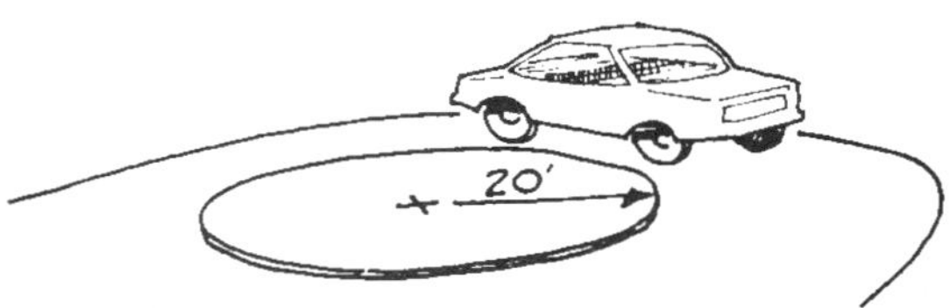

Figure 4-12
Tight turns are hard to negotiate.

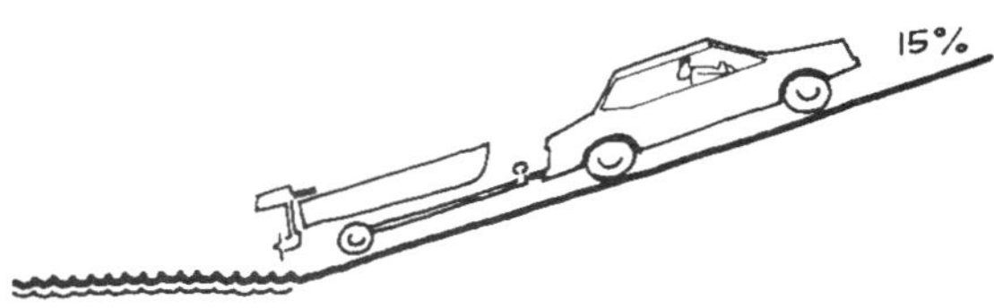

Figure 4-13
Steep pitches are tough to manage.

Figure 4-14
Drastic slopes are difficult to mow.

facility after construction. A cooperative spirit benefits both parties here. Although an understanding of the operator's methods may point one way to efficiency, alternative and possibly more efficient modes of operation might be offered to the administrator by the landscape architect as design possibilities unfold.

Matter of Concern

6E. Seats and Seating

Machines and facilities operate. Well, people also "operate." Their physical limitations must be appreciated in design especially where comfort can be affected, as demonstrated in Figure 4-15. Does the edge of the seat catch your leg at mid-calf or mid-thigh, or does seat height allow your knee joint to curve comfortably? Does the slant of the seat cause your bottom parts to slide unceremoniously into the gap at the base of the back slats? Are the back slats spaced so that each digs into your spinal column? Comfort also decides the question of whether to have a seat back or not to have one. A backless bench may suffice for transient sitting, like waiting a few minutes for a bus, but comfortable back support is essential for a long lunch break or for reading the *Chicago Tribune*. Try every one you see and note the manufacturers. Remember the good ones. Bench standards are included in Harris and Dines' *Graphic Standards for Landscape Architects*.*

One thing all seats must have in common is the height to the plane the human rump encounters. All seats have to be at knee height (14 to 18 inches accommodates the widest variety of humans).

Assuming the *bench* is comfortable, what about the bench*es*? Sitting on the world's most comfortable bench (right back slant, good seat form, perfect height) can still be very uncomfortable if it is isolated or placed back to back with another bench.

People like to watch the action. They enjoy talking across easy distances. Effective placing of benches can provide another important level of comfort by embracing this need—by grouping, facing, and clustering benches to create spaces off the routes where people walk (but not necessarily out of sight of them). (See Figure 4-16.)

Seating. Is this really important? Considering that the most significant activity in any park is **sitting** (ranking just above **watching**), then *yes*, it's very important. People are very sensitive to how they sit, where they sit, and who they sit near.

*Charles W. Harris and Nicholas T. Dines (1998). *Time Saver Standards for Landscape Architecture*. New York: McGraw-Hill, pp. 340–349.

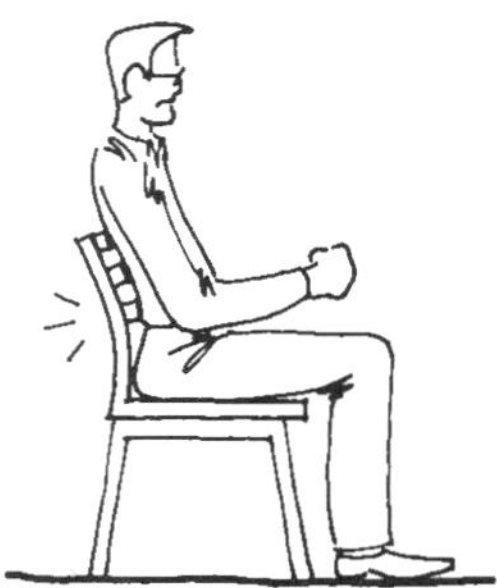

Figure 4-15
Bench design must consider the physical limitations of the person.

To offer guidelines supporting this very important activity, we will deal with *seating*, not just seats. In a series of experiments, we observed some groups of concrete slab benches, all of which were 30 inches in depth and 6 or 8 feet long. The benches were situated in busy places on a campus in the Midwest. The even lengths divide into 24-inch spaces, slightly more than the average size of hairs of all kinds. So the 6-foot benches *could* accommodate three people, the 8-footers easily four people.

In the first observation (Figure 4-17 on p. 54), three 6-foot benches at a bus stop are arranged in an L shape. The first person sits at end of a bench, as do persons two and three (stage 1). Persons four, five, and six sit on remaining ends of benches (stage 2—note that person six took the last end, at the corner of the L). The next person has to decide which persons to sit between . . . or not to sit. Some delay. Several people stand rather than accept the nearness of others, but the benches do fill up (stage 3).

In the second observation (Figure 4-18 on p. 54), four 8-foot benches at another bus stop are arrayed in a line. Over a longer period, several mixtures occur. Ends were sometimes taken on two separate benches. The next user would frequently sit in a middle position. This would really screw things up. The middle sitter would be alone while people would sit between two end-sitters. As a crowd built up, the benches would fill with three, but very seldom with four unless they were clearly friends.

In the third observation (Figure 4-19 on p. 54), still at the 8-foot bench bus stop, we tampered with the benches early in the morning, putting tape lines (or "markers") at 24-inch intervals along the benches, clearly marking off the (adequate) individual spaces. As before, the ends filled up easily. The inside spaces were now a problem. Several people would stand rather than sitting in the marked space. People did take the spaces eventually. But *nobody* touched a tape line! There just as well could have been a sword blade marking that line.

People by nature, at least in this culture, are reluctant to impinge on someone else's space, and they are also very sensitive to rules even if they're only hints—which makes very important the design details of the seating offered to the park users who do want to sit comfortably, and definitely properly.

Encyclopedia of Seating. Take a moment to explore the Figures 4-20A and 4-20B (on pp. 55–56), the Encyclopedia of Seating. Imagine as you review the various situations and relationships that each specific "person space" could be a chair-seat-module with a back and any kind of actual surface material. The details are important. But it's the relationships that have to work. In reference to basic distances, we recommend, with acknowledgment of a lot of variations, that *face-to-face* lines of seating should be at least 6′ apart; and *corner-to-corner* distances need to be about 4′ apart.

How do people walk? In the outdoors, long strides are normal. To maintain a fluid pace, risers for outdoor steps must be comparatively short and treads relatively long. In the interests of walking comfort, landscape architect Thomas Church suggested the following ratio: twice the riser plus the tread equals 26 (Figure 4-21 on p. 57). Thus, if a riser is 5½ inches, the tread should be 15 inches; where the riser is planned at 6 inches, the tread would be 14 inches.

These dimensions are very real in terms of human anatomy and range of motion. Approaching a courts building in a major city in Florida, the pedestrian ascends a long flight of steps to the entry. The

Figure 4-16
Bench placement can foster or frustrate interaction.

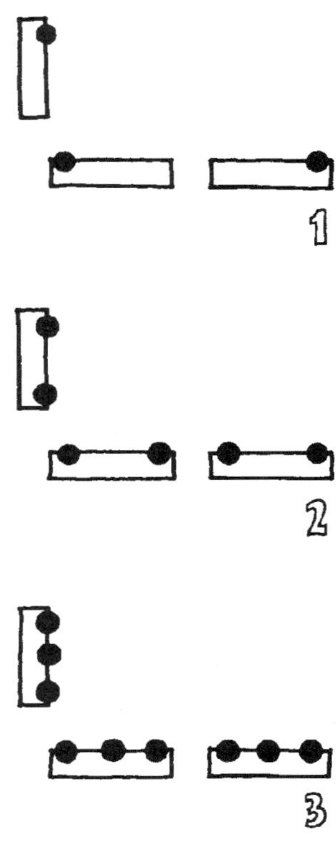

Figure 4-17
The "bench observation" set-up.

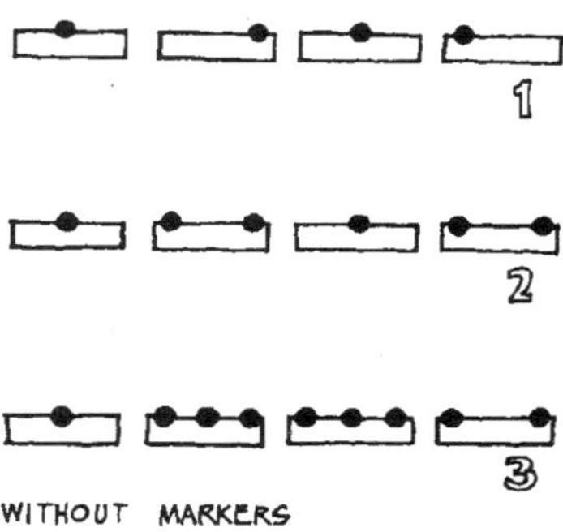

Figure 4-18
Unmarked bench sequence.

Figure 4-19
Marked bench final pattern.

designer, apparently trying to be kind and reduce the effort, had set the treads at a normal depth, about 12 inches, and had reduced the risers to a bit more than 3 inches. Sounds insignificant. But groups (or individuals) came to these steps in conversation or in contemplation of legal affairs, clearly not paying any attention to the actual steps. Almost invariably the travelers would make no more than three steps before the feet unconsciously reached for a tread and were attacked by a riser. Stumbles resulted, occasionally a fall. This seems like the worst place to fail to heed the safety implication of proper step design, considering the litigious nature of the passing population!

Where more than one foot must be placed on a tread, as in pedestrian step-ramps, tread length should be such that alternate legs can be used in the stepping-up process (Figure 4-22 on p. 57). To have to step up on the same foot all the time is quite tiring.

Height considerations apply as much to a street curb as to a step riser. Our constant experience and consequent expectation is of a 6-inch change of grade at a curb. (See Figure 4-23 on p. 57.) Much higher or lower spells disaster. Sighted or blind, we expect the 6 inches and will trip over a curb of a different height. Curbs are understood to define traffic separations and are frequently used in protective patterns to enclose planter beds, sculptures, or other special areas. The type of edge definer that is higher than a curb but too low for a seat verges on useless. If the design sense of an area calls for more than a conventional curb-size edge, the plan should at minimum call for a "sittable" 14 to 15 inches, so that the added material also equals added usefulness.

Similar insights should govern fireplace and drinking-fountain heights, as well as point the way to such appreciated conveniences as ramps integrated into stairways for pushing baby carriages. (See Figure 4-24 on p. 58.) Comfort should be a hallmark of every design.

The impact of federal legislation mandating accessibility has been so great that at the mention of ramps, most persons think of wheelchairs. We should give a bit more attention to the nature of this legislation while considering human operating needs. The law was developed around the eminently sensible concept of ensuring access for all people to all public places. Although persons in wheelchairs happen to have highly visible access limitations, the law also benefits those with artificial limbs, degenerative diseases, braces, or arthritic ailments, pregnant women, the person with a broken leg, or people suffering from *any* condition that limits mobility. At some point, that's *all* of us.

Under the legislation, the standard maximum slope for ramps is 8.33 percent, equal to a ratio of a 1-inch rise (vertically) to a 12-inch run (horizontally). The ramps must include a handrail. This is a maximum. The ratio can be less—a point frequently misunderstood by designers. In fact, if the slope is 5% or less (a proportion of 1 to 20), no rail is required. Given the choice, what slope better satisfies the operating need for accessibility? The 5% solution comfortably accommodates all persons, meets accessibility requirements, is good in all weather (including icy conditions in northern climates) and, advancing our next principle, is less expensive since no handrail is required. "Take it to the limit" is only a song lyric; it isn't a design directive.

SELECTED SEAT SHAPES

Showing user orientation limitations and potentials

STRAIGHT SLAB

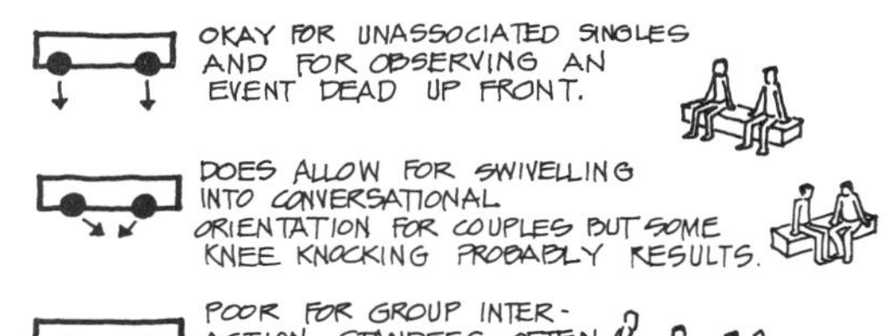

SINGLE POD

OKAY FOR SINGLE OCCUPANT OR (DEPENDING UPON SIZE) 2-4 UNASSOCIATED SINGLES FOR BY PERMITTING BACK-TO-BACK SITTING, USERS MAY BE ABLE TO "TUNE OUT" THE OTHERS.

POOR FOR COUPLE INTERACTION BECAUSE OF EITHER SIZE LIMITATIONS OR HINDRANCE POSED FOR EASY SWIVELLING; POOREST FOR GROUP INTERACTION.
LONGER UNITS TAKE ON THE CHARACTERISTICS OF STRAIGHT SLABS.

SINGLE JOGS

MULTI-JOGS

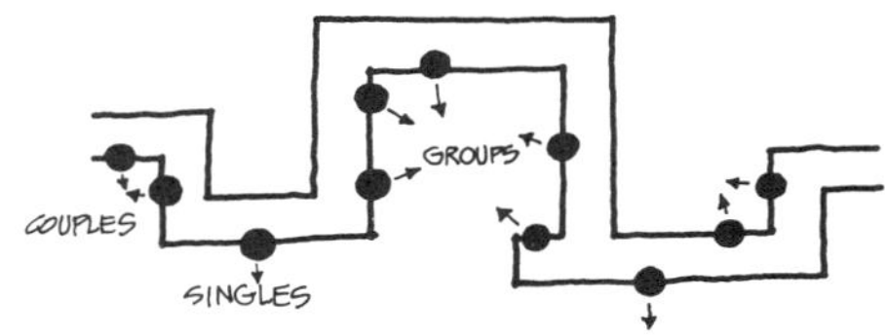

BEST: ACCOMMODATES A VARIETY OF DEMANDS.

CIRCLE

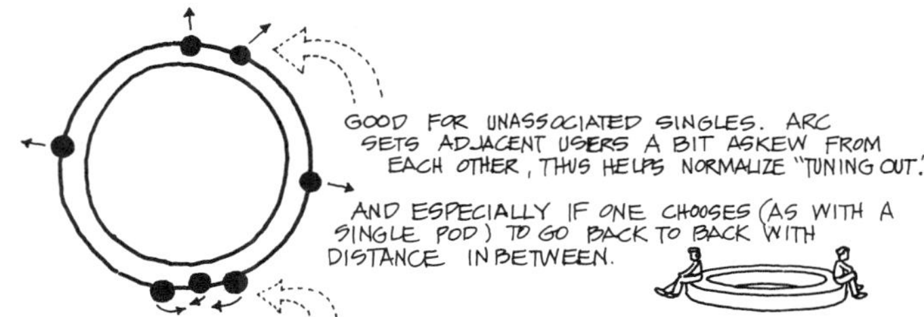

CONVERSATIONAL ORIENTATION POSSIBLE AMONG COUPLES BUT SINCE THEY MUST SWIVEL AGAINST THE SHAPE, IS MORE DISCOMFORTING THAN STRAIGHT SLAB; POORER YET FOR THIRD PARTY WHO MUST BALANCE ON ONE BUTTOCK TO REMAIN IN THE ACT (THE TIGHTER THE RADIUS, THE GREATER THE PROBLEMS); JUST AS BAD FOR GROUP INTERACTION AS STRAIGHT SLAB.

CURVE

(LIMITED TO CONCAVE OR CONVEX POTENTIALS-LIMITATIONS IF ONLY ONE SIDE CAN BE USED. NOTE POSSIBILITIES, THOUGH, IF SITTING IS ALLOWABLE ON BOTH SIDES AS SHOWN.)

CONVEX SIDE (AS WITH CIRCLES) OKAY FOR UNASSOCIATED SINGLES BUT POSES PROBLEMS FOR CONVERSATIONALISTS.

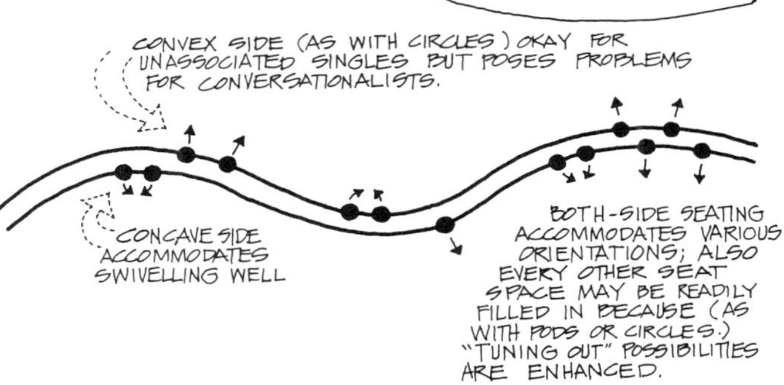

Figure 4-20A
Encyclopedia of seating: Selected seat shapes.

SELECTED SEATING ARRANGEMENTS

Showing some critical distances to be kept in mind (|⟷|)

Where a complexity of use and users is anticipated, strive to provide great variety in shapes and arrangements.

STRICT LINEARITY

(DUE TO LIMITATIONS POSED FOR CONVERSATIONALISTS, ESPECIALLY WHEN GATHERED IN SMALL GROUPS, AVOID THIS AS THE EXCLUSIVE ARRANGEMENT AT ALL COSTS.)

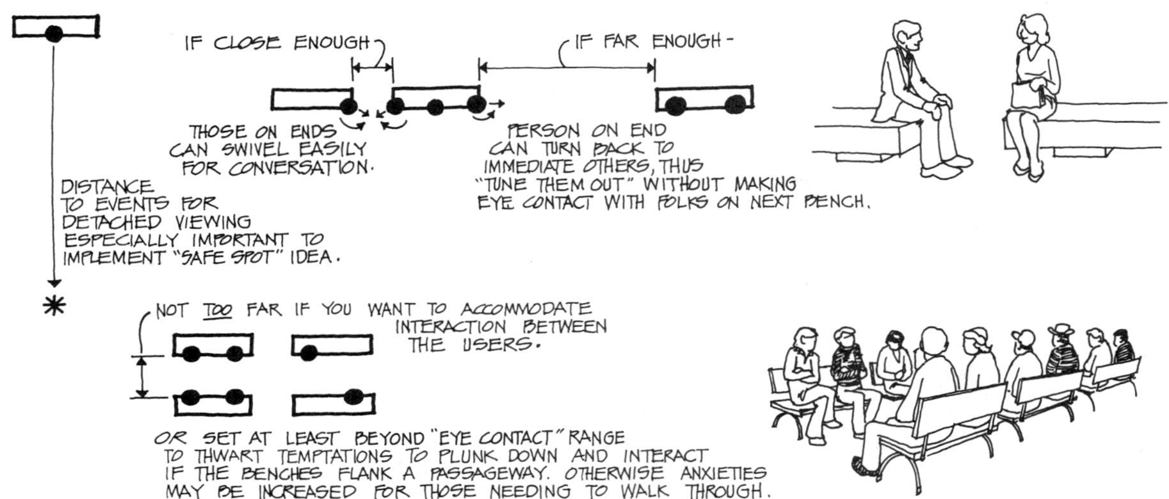

RIGHT ANGLES

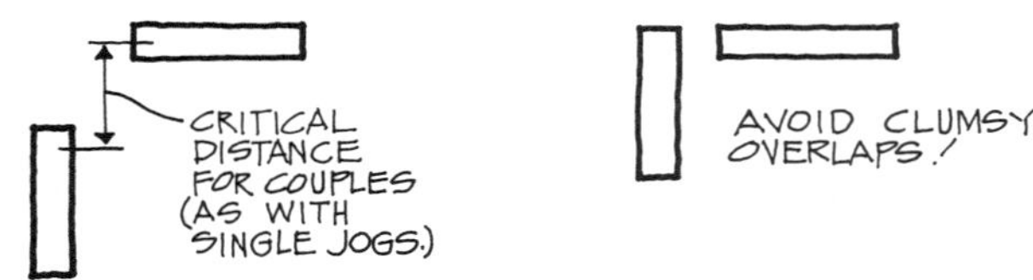

CLUSTERS

Figure 4-20B
Encyclopedia of seating: Selected seating arrangements.

Principle 7: Meet Needs for Lowest Possible Cost

Although technical needs are rather easy to identify, satisfying these needs as well as meeting more elusive requirements is often balked due to lack of funds. Here is the most classic conflict faced by the designer, one for which he or she must establish a balanced set of priorities: the conflict between needs and budget restrictions.

It is incumbent on designers to avoid unnecessary costs. They must suggest only what can be supported by sound purpose. Yet, in another sense, they are obliged not to skimp, for professionalism dictates that their design must satisfy the true needs of the development. The only way out of this dilemma is through a partnership agreement between the client who is footing the bill and the designer who is assuaging a professional conscience. They must *both* determine what the needs are. And they must *together* ensure that their list of needs is compatible with the monies available to construct the project.

Also, as mentioned earlier, the present decrease in resources (both land and money) forces the designer to consider all proposed work in terms of increased productivity from the resources committed. Seasonal use of facilities, use only by restricted populations, use for exotic purposes exclusively—all of these suggest a lack of consideration of this increasingly important aspect of design.

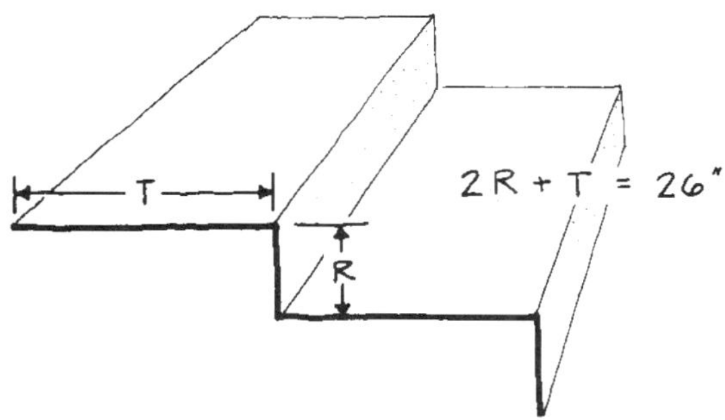

Figure 4-21
Ideal proportions for outdoor steps.

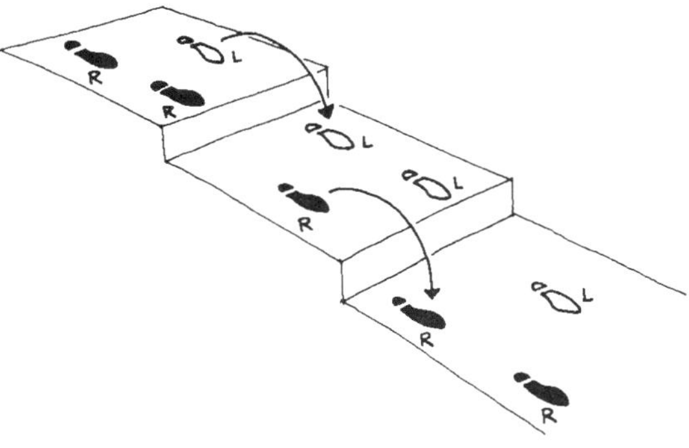

Figure 4-22
Step-ramps should be designed to allow the user to negotiate the risers with alternate legs.

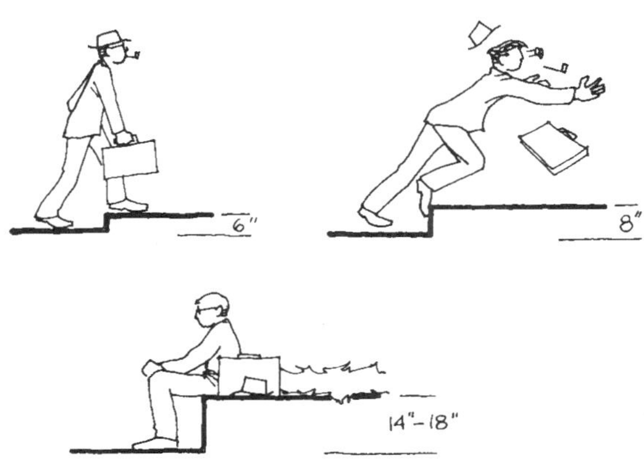

Figure 4-23
Curbs can be confusing or comfortable depending on height.

Matters of Concern

7A. Balance of Needs and Budget

A meeting of the minds between client and designer begins at the onset of design thinking, when an initial program of directives is developed in a brainstorming session attended by the landscape architect and the owners or administrators of the project. It continues through preliminary stages, when the designer comes back to the client with early ideas that might include objectives not apparent at the time of program development but revealed during periods of design research. And it follows throughout the final stages, when the work is let for bid among contractors and when changes are being contemplated during construction. A proper balance among needs and budget can be most readily struck if budget is discussed at all such gatherings. How much money is available? What are the alternative ways of meeting goals under financial stringencies? What are the high- and low-priority items? What can be constructed in stages to spread out construction costs over a number of years?

In addition to construction funding, the type, extent, and finances of maintenance will dictate a great many design decisions and therefore deserve an important place in the discussion. It is as foolish to consider exotic plantings if there are no expert horticulturists on the agency's staff as it is to plan for a public swimming pool if there are no funds available for staffing and upkeep.

During consultations dealing with program, it is usually the responsibility of the designer to draw a full expression of needs out of the client and add to them as experience suggests. But initiative for identifying needs can come from either party. Regardless of who is the prime contributor in this regard, the key to success of the partnership in the long run is agreement before final design commitments are made. If this trick is turned, each party is forewarned of minimum

Figure 4-24
Consider the comfort of the user in the design of outdoor facilities.

needs, their relationship to the requirements of budget, and what must be done in the realm of compromise in order to turn on-paper design into reality.

While designer and client agonize over the restrictions of the budget, the goals it won't reach, and the stretching exercises painfully in process, it is worth putting a last item on the agenda prompted by the pervasive key word *why*. Are these proper goals? Is this Sisyphean trip necessary? Sometimes it may not be.

One park-school complex fought with a boggy area created by outflow drainage of tile lines from several surrounding sites. The ongoing effort was to try to establish turf consistent with the rest of the property while trying annually—and vainly—to set aside enough money to permanently drain the area. Neither fight was being won. At the conclusion of a fortunately quiet meeting, the problem was brought into focus. What was the prize being sought? If drained and conventionalized, what was this area to become? It was on the periphery of the property, not within the area programmed for high activity. It was not needed to supplement other space-restricted facilities. By being "unused," it was being used: casual strolling, play, idle curiosity comprised its program. What would it do if nothing further was done to it? It did what it had been trying to do: grow into an interesting little nature preserve which, as its species of plants and birds approached a state of stable self-care, became a useful resource for school and park nature programs as well as a point of pleasant contrast to the suburban surround. Best of all the benefits was the price: nothing.

The cities of Chicago and Boston have proceeded with some success to incorporate "urban wilds" in areas associated with water bodies in central city parks. Formerly manicured (expensively), the areas were left or biologically encouraged to return to a shaggier wildness to encourage the development of habitat in which animals and birds would exist successfully and be seen in their natural settings. Citizens have been very supportive, and the sites have become educational resources.

The point to be made is that designer and client must both keep an open mind on an apparently closed issue. Good planning has to be flexible planning, with constant effort being given to turning problems into opportunities. Remember the key word *why* and the key question "What else *could* it be?"

Jerrold Soesbe, from his experience as chief landscape architect for the Lake County Forest Preserve in Illinois, provided an example of the problem-opportunity sequence resulting from a budget squeeze.

Acre upon acre of rolling turf at the preserve was a mowing headache in terms of schedule hassles, personnel commitment, and fuel cost for equipment operation. Yet the burden was carried because the clientele, thousands of suburban Chicagoans, cared a lot about the area, used it intensively, and were definitely not bashful in their response to changes. But something had to give. The giving, which proved to be a real gift, took the form of a decision to recognize the value of the original prairie of the area as a historic and ecologically appropriate condition. The next decisions made were to move the turf back in the direction of that prairie by ceasing the mowing and identifying the area as a special place. Subsequently named the Old School Preserve, its special prairie status was further reinforced by a long-term planting program incorporating additional prairie grasses and plants.

The total success of the solution seemed obvious, but also obvious was the need for some very effective public information work. Someone had to explain to the thousands of constituents who had expressed their admiration for the grand sweeps of trimmed grass that the love affair was over. Appreciating the sincerity and depth of public emotion, the staff developed a careful, creative, well-researched program to provide education about the concept in practice, its long-term benefits and beauty, the budget considerations—and never needed to exercise any of the effort. Even before the campaign was launched, the visual evidence of softness in the unmowed tracts began to get highly favorable reactions from visitors. The right time and right target resulted in a painless and very successful transition.

The basis for success in this situation was the straightforward recognition of a problem presumably requiring a more-money solution followed by a detour around that solution into opportunity. Although it might seem sheer luck that they were favored by a totally in-tune public, the staff's knowledge of that public and consequent preparation was solid, and they were ready to win their clientele over by the clear demonstration that a limitation had been converted to an asset. The award-winning project has become a major attraction for the region. Before the development, nobody *needed* a prairie restoration, at least not in the sense that someone had identified that need based on some knowledge from or of the client. The need was for a solution of a budget problem. The product was the result of the landscape architect's creative view of the whole situation.

Whether on the desk, on the lap, or in the pocket, the computer as business companion is a gateway to worldwide information on the Web, with access to data allowing great improvement in balancing needs and budget, particularly when the issue is maintenance. The prairie conversion mentioned didn't start in that direction; the problem was the cost of current maintenance practices. As a major budget highlight, this one was easy to spot even though the solution required some major imaginative effort.

In a more metropolitan system, however, with a broad variety of park types and even greater variety of maintenance activities, the sheer number of separate services performed makes conventional analysis very difficult. Can a comparison be made of grass cut here to grass cut there? (What about paving repairs or upkeep costs in different areas? Replacement costs for various types of equipment used in a number of different locations?) Of course it can.

Broadscale analysis of maintenance has been done with commendable success in many situations. An exponentially growing number of digital records are increasingly and more easily stored, and overlay comparisons of data are more easily made. Using grass mowing as an example, one would consider the complexity of the mowing operation in an area (few obstacles, simple; many obstacles, complex), type of equipment used, slope of ground, type of use, public prominence (high or low exposure), and costs of labor associated with each area (dollars per square unit). Using the computer's capability of lightning-speed data review, areas with strong apparent similarities can be compared to determine whether maintenance is cost effective. If Area A has conditions seemingly identical to those in Area B yet costs 20% more to maintain, a closer look is appropriate. The importance to both the designer and the user is that each can intensify this search until either a flaw in operations is isolated or a design problem is spotlighted, allowing a better design solution to be developed. The happiest result is the improvement of an environment physically and fiscally.

Beyond the conventional procedures for project development, as "creative" financing has become a standard term in economic parlance, so *creative development* should now mean more than imaginative design. The creative alternatives that deserve consideration are those that offer increased financial capacity and expanded activity opportunities. The possibility of joint public-private financing is one such opportunity. The sharing of resources expands what each party has to offer and benefits both. The park agency, for example, has the property and definitely the audience; the private developer of, for example, a waterslide facility has the expertise and experience to create an exciting and profitable project. Yet neither has the means to succeed without a partner.

Results of this kind of cooperative venture have been successful in varying degrees across the country. The only drawback has been the initial shock experienced by neighbors of the new development. A conscientious park agency can eliminate many problems simply by open-eyed preparation of the residents affected and a very thorough, protection-minded investigation of the private developer (accompanied by a carefully prepared contract) before a pact is sealed.

Boston's Post Office Square is a highly successful public space over an existing garage, an example of private-sector leadership on private property with public parks cooperation in development and maintenance. As discussed in the August 2001 *Landscape Architecture Magazine* ("In Search of Public Space") Post Office Square is a green, clean, pleasant area that becomes packed every noon hour. Maintenance is actually by contract, paid by private funds from parking revenue.

Partnerships between public and private sectors don't need to be limited to those in which the private partner is the developer. One extraordinary example of financial partnership features the private parties as the funding source and the park agency as the developer. The little town of West Lafayette, Indiana, needed a new roof on the city pool's bathhouse. Estimate: $60,000 to $70,000. The city simply didn't have the bucks. They didn't even really have the money to pay the heating bill to keep the pool water warm.

After some amazing brainstorming sessions, finally topped by some revisions to state legislation to permit the action, the city actually arranged with a group of private investors to let them reroof the bathhouse, incorporating a complete solar water-heating system. The investors lease the roof space from the agency for $1 a year, then sell the collected energy back to the agency at a set rate. (The rate can be renegotiated at the end of the nine-year contract.) The investors even include maintenance of the system in their offering. Their gain is in the very healthy return from the energy sale and the considerable tax benefits from their investment. The agency got a new roof in addition to a new heating system at an excellent price. The system is so productive that during seasons when the pool is not in use, the heat is rerouted to a nearby junior high school, where it heats the showers and lavatories.

In the span of a decade, the further appreciation of the benefits of roofs becoming green *growing* spaces has expanded rapidly. The result has been a rapid increase in roof gardens covering much of the roof space on flat-topped buildings common to offices, maintenance services, storage facilities, and other uses. Evidence shows that planted roofs result in a reduction in heat loss from the building and the reduction of summer heat buildup. The transition has proved to be a major cost saving, more than adequate to justify the expense of the installations. And they look great.

Apart from the possibility of enlisting private developers as partners, a park agency has the built-in potential of joint ventures with its own citizens: programs in which citizen groups take responsibility for elements of the parks system in which they have a special interest. Neighborhood parks are examples of properties that residents may recognize as of significant value to them and feel enough concern to participate in development programs. In one agency, neighborhood people took on the responsibility for nearly all of the maintenance of their park in a contract-type relationship. This "Adopt-A-Park" program saved the agency a considerable amount of money and time, as well as increasing the local pride and interest in the park and its programs.

Other examples more commonly encountered but equally important in terms of economic benefits realized include special-project efforts by civic groups—a "tot lot" constructed by a local business club, a picnic area cleaned up by a scout troop, or a fund-raising sale by a parents' group to provide money for the park agency to make some improvement that would otherwise be unaffordable. In every case, one of the major benefits in addition to the project itself is the pride and even in some cases improved security of the area resulting from the increased attention generated.

7B. Use of Existing Site Resources

It is naive to suggest that extensive projects can be funded for peanuts. You get what you pay for. But the alert designer takes pains to satisfy needs at the lowest possible cost. Paramount is the incorporation of existing site resources into the plan. The designer tries to make best use of what is already there.

The main point here has already been presented in the section on relation of use areas to site; that is, facilities should be assigned only to portions of the site that are compatible with that use. We reemphasize this because it is so often the key factor underlying the success or failure of land-use projects. Implications range from the most obvious—camping can be disappointing if engaged in upon a poorly drained soil and expensive to accommodate if the land must be underdrained—to the less apparent—conifers (especially pines) die quickly where there is concentrated foot traffic, whereas such hardwoods as hickories and sycamores seem to withstand such concentrations relatively well. Hence, certain species must be ruled out for high-impact uses while other reasonable candidates are well suited to that purpose. (See related tables in the appendices.)

We have also said that use areas should be placed upon portions of the site which, in their unimproved states, approximate the desired finish grade: football fields in the existing flats, toboggan runs on the available north-facing slopes, for example. Doing this satisfies two concerns. It minimizes the cost of earthmoving and simplifies the blending of the new with the old. Not only does such attention to function and aesthetics in the same move support the contention that neither can be separated in design thinking, but it also suggests that the inclusion of aesthetic considerations does not necessarily mean greater expense. Indeed, in this instance, money can be saved.

We noted previously that site modification to provide access for persons with mobility limitations were unaesthetic if developed as a tacked-on afterthought. This is an abuse of existing site resources. Let's look at an example of a good use of an existing site resource for access. Tim Nugent, retired director of the University of Illinois's renowned Rehabilitation Center, has described with pride the planning change made by a large church. In the process of designing a new building, they had planned a generous parking lot in front; a spacious entry porch elevated a few steps above the grade of the lot; and to accommodate their mobility-limited members, a ramp had been proposed winding back and forth from the parking lot to the porch, thereby providing the common point of entry for everyone. Tim's questions about the plan led to the removal of the ramp and the revision of the planned grade of the entire parking lot (still close to prevailing grades on the site) so that the lot itself slopes up to the original porch grade, eliminating the need for any steps up to the entry. Access for everybody was simplified. The building gained rather than lost in overall appearance. And the cost? Cheaper. (Same amount of paving for the lot, but less grade modification, less material, and less labor since there was no porch or steps to form.)

Savings can also result from identifying suitable development areas based not on their "natural" conditions (slopes, vegetation, soils) but on their present declined use being "transitioned" into a more productive new use. A train station sitting by an abandoned rail line has excellent potential as a picnic shelter, especially if the vestigial adjacent paving can be reclaimed for parking and access. The railroad right-of-way itself is an asset convertible to a hiking trail or bikeway with excellent, easy grades for circulation and frequently with interesting, even historical, plant communities associated with it (Figure 4-25).

When landforms are flattened to provide field-game areas, slopes between use areas may be a necessary by-product. Even though they are too steep to serve a conventional function, the designer, in the search for additional uses, might identify these slopes as appropriate, safe places to locate children's slides or sled runs. More about these later.

Studies conducted for a developer of golf-course-based communities were prompted by a realization of, on one hand, an expressed need for more compact housing as duplexes or apartment condominium units. On the other hand, the golf courses were showing a reduced usage in favor of shorter, tighter (easier) holes for many users. Much interest was expressed in multiuse athletic centers closer to the user population and available for the all-day times rather than the daylight limits of the golf facilities (Figure 4-26).

Figure 4-25
Transition of a railroad track through decline, removal, and renovation to a trail.

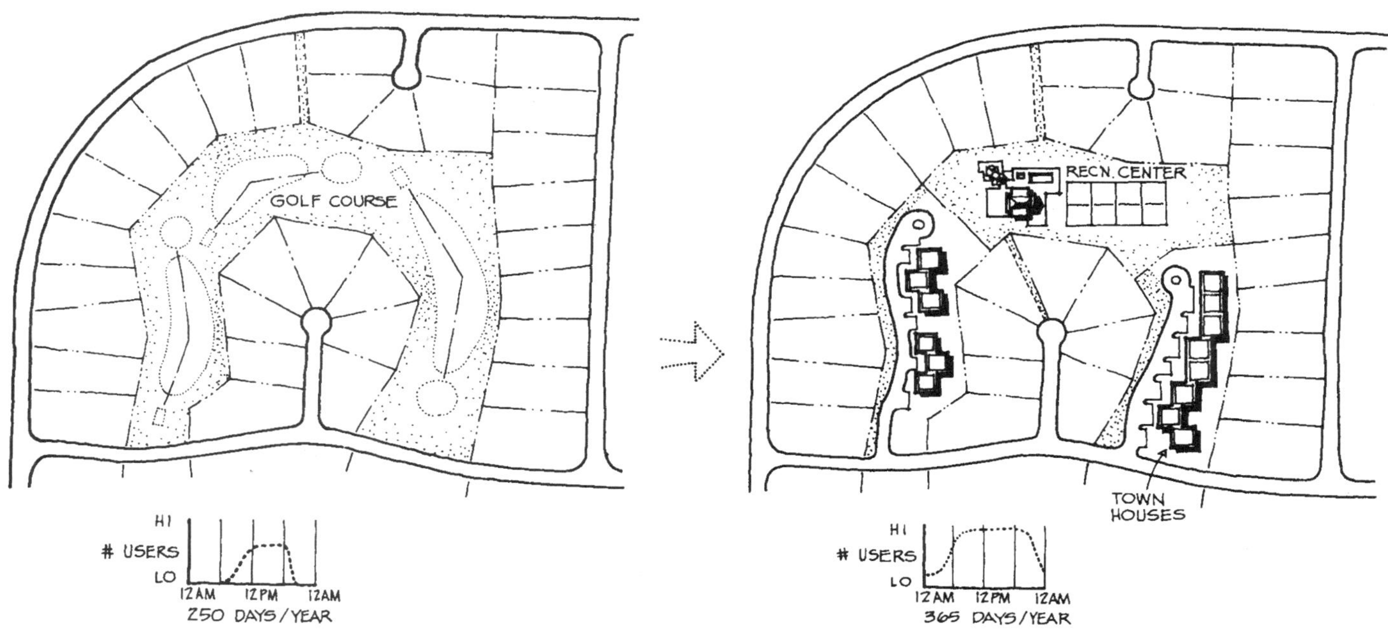

Figure 4-26
What else could it be? A housing-golf complex becomes a more productive recreation center.

The study resulted in variations of course redesigns, eliminating some holes completely in favor of adding new, small housing units as well as compact recreation centers in the same space. The result would be much greater usage and a considerable economic gain in the housing sales and the facility fees made possible.

Before one agency in North Carolina establishes a program for the development of a site, they develop a team of non-typical experts to ask the park what it *could* be. Soils, water movement, vegetation, landforms—that's the usual stuff. The investigating experts are there to ask the site about its archaeology, its history, its folklore, its influence on its neighbors. The result of such inquiry has been a variety of surprises such as ancient trails, spectacular views with special history, surprising animal behavior, and things generally unmapped and unknown without the special effort. The ultimate result has sometimes been a park by preservation and protection more than by development, where the park is for education and entertainment rather than for the typical activities.

Why? Why not? This is not a recommendation to do nothing in reaction to an apparent problem. The point is to make sure to really look at what each site offers as opportunity in contrast to the possible problem.

A designer who recognizes the natural forms of the land as a resource is listening to a site's advice. Based on good listening, a case can be made, for example, for locating roadways in the available linear spaces created by earth forms (Figure 4-27 on p. 64). Since such spaces are usually part of the site's natural drainage pattern, both the aesthetic advantages of enclosure and the functional benefits of eliminating expensive drainage channeling are gained. As shown in Figure 4-28 (on p. 64), this is decidedly true where buildings and major use areas have been located on high ground. Following the drainage network of the site, rainwater is thereby encouraged to continue its natural course, running away from the buildings, down the slopes into the

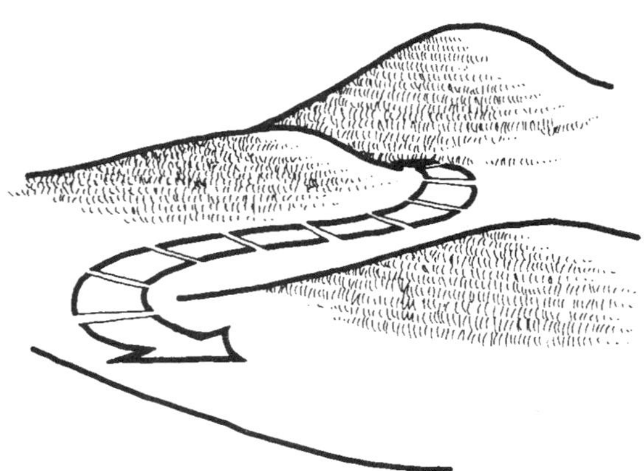

Figure 4-27
Valley floors are natural drainageways.

Figure 4-28
Roadways placed in the valleys readily take away water draining from use areas located at higher elevations.

roadways, and thence off the site. The site itself, not the (expensive) bulldozer, has done most of the work for you.

Consider other possibilities for checking construction costs. How about using existing buildings to complement the new structures? Renovate the barn into a maintenance facility or put it to use as an arts and crafts center. Slap some paint on the tool shed and it will serve as a control center for the tennis court battery.

The Indianapolis City Department of Parks & Recreation recognized potential in an existing site resource. One park contained a huge old swimming pool developed for municipal programs of the post-World War II period. It was vast, it was deep, it was dangerous, it was hard to maintain—and it was practically never used. The agency's director of research and development noted that the pool might have some value as structural support for a smaller and shallower intensive-use pool, which was subsequently developed and has been very successful. Peripheral deck paving and other supporting facilities have been built using other remnants of the original pool as a base.

What can be done to eliminate the expense of special footings? See that structures are kept off unstable soil and are placed upon naturally solid bases. Why not eliminate the cost of tree removal and the expense of new plantings? Wrap new structures and plantings around existing trees. Place facilities where desired grade already exists so as to minimize the possibility of bulldozer damage. Put facilities enhanced by plantings in places where plants already exist; place open-expanse facilities in spaces uninhabited by tree groves. Remember that it will take at least 15 years for a newly planted sapling to produce substantial shade and effect, whereas the existing tree might already be of adequate size. How can you ease the financial burden imposed by the construction of lengthy gas, water, electrical, and sewage lines? Locate heavy utility users near existing trunk lines or, in the cases of sewage and water, adjacent to soil suitable for disposal or drilling, thus cutting down the length of feeder pipes and cables.

These and similar measures for taking advantage of what already exists are decided upon in the design stage of development. After design, the advantages are lost. There is usually little excuse for such loss except downright nearsightedness. And that's only a short step from incompetence.

7C. Increased Productivity

The plan at which one arrives by putting all the existing resources of a park site to their most appropriate use is, in a sense, a two-dimensional solution. It becomes three-dimensional when one takes advantage of overlap potentials to create layers of use occupying essentially the same space. For example, the outfields of baseball diamonds can serve as football and soccer fields, or the roof of a parking structure can serve as a tennis court. The ultimate increase in the productive value of a site has to reach into the fourth dimension—time. Analyzing how often as well as how a site is used is as important as knowing its soil type.

Defining a paved area as a service court, for example, may be a solid choice in respect to its location (adjacent to the recreation center it supports, good connection to the primary circulation, but out of sight of most traffic) and the details of its design (proper dimensions for turning movements of trucks, proper slope for drainage, appropriate thickness and base strength for vehicle loads). But the designer in this case has not factored in the fourth dimension. The area's use times turn out to be fairly specific: intensive early morning start-of-duty action, slight activity at noon, then heavy again at the end-of-duty late afternoon period. The rest of the time—mornings, afternoons, evenings—practically no action. What else could occur on that surface in that location and complete the four-dimensional plan for the space? A natural choice would seem to be an activity like basketball, which could be added for the minor cost of a backboard and possibly some minor line additions. It may not be an official Olympic layout, but most of the play and most of the fun is found in the most casual circumstances. That's what a pickup game is all about.

Other use innovations may be serendipitous responses to recurring problems. At a Chicago university campus, utility repairs and changes were a constant grief, resulting in what was frequently referred to as the "Trees on Wheels" program. To clear areas for utility or other construction work, trees had to be yanked out of the way, sometimes on uncomfortably short notice. In the best of situations, the trees would be relocated to the last work area. In the worst situations, with no place to put them, they had to be chopped. Recognizing that an adjacent playfield area (about a block square) represented a reasonably flexible site location, planners decided to adjust the south side of the combined baseball and football field to permit a line of trees to be set inside the utility right-of-way line bordering the property. In this strip, the refugee trees were placed as in a nursery (Figure 4-29). With no loss of function, the playfield gained a spectator shade band, a definite benefit even though none of the trees would necessarily be permanent and the density of the band would frequently change as some of the trees were relocated onto the main campus. The biggest gain was a reduction—in construction-sacrificed trees.

Figure 4-29
Nursery for the "Trees on Wheels" refugees.

A park agency in Texas realized that dikes forming a detention basin could be designed to become the amphitheater seating areas for ball games and musical performances staged on the open grass floor of the basin (in dry times, of course). In the Midwest, another park agency and a local university shared use of a huge fill mound on university property. Fairly remote from active urban areas, the mound changed form constantly as construction for both agencies added or subtracted material. The park agency's security personnel struggled in vain to keep dirt bikers off the mound. Finally, the park agency turned a problem into an opportunity by programming successful events on the hills. Conflicts between construction traffic and trespassers substantially disappeared as biker groups recognized scheduled events and supported the effort through their own membership.

Accommodating new recreation in the form of specialized vehicles like dune buggies, dirt bikes, snowmobiles, all-terrain juggernauts, or even non-terrain hang gliders means that more users are served, and this means increased productivity. Many of these specialized vehicles can utilize property previously considered fallow ground (too steep, too wet, too inaccessible). However, the designer must carefully appraise the impact of the new use on existing conditions so that the new activity doesn't destroy precious plant or animal life prized by other users for other reasons.

Suppose you discover an area suitable for two or more different uses but realize that the extra functions can't be accommodated because of time conflicts. This need not be a dead end for a good idea. By investigating each activity and its apparent time requirements you may identify some possible changes in scheduling that will allow for increased use. Recognizing the need for additional parking for certain use areas might suggest the building of a parking lot. But a better solution would be to negotiate with the adjacent church for use of their lot at times when the church is unoccupied. The cost is only the rescheduling of events and possibly providing improved circulation between the park and church.

Faced with some unpleasant bullet-biting after some historically severe revenue reduction legislation made history, the state park system of California recognized areas of potential productivity gain and made maximum use of its resources. Roadside parking areas with regulations preventing use from sundown to sunrise became overnight camping areas. Campers could park recreational vehicles from 7 PM to 7 AM. The result was a new service for campers on the move during the day, provided without infringing on the daytime visitors who need the parking for their destination-oriented use. The number of satisfied customers was doubled.

Other bonus discoveries require simply a closer look at how space is used. The edges of playfields and the spaces between them may be correctly plotted on the master plan and accordingly constructed on the site but, as many planners have discovered, there is still good space available for threading a jogging trail through these tracts. Even better, exercise facilities can be appended to the trail; little pockets of extra space can be used for the small but high-intensity equipment for chin-ups, sit-ups, stretching, and other specialties for muscular development.

Low-intensity use of field facilities is generally the result of protective concern rather than poor scheduling. Turf doesn't wear well. If it gets bruised in a rugged game, it ordinarily has to be given compensating "R-and-R time" for reseeding and recuperation. In a dream scenario, a field heavily used on a given day is taken out of service for a while so new grass can spread back into the damaged areas; play continues on the alternate fields. Sure. Everybody has extra fields lying around—they are kept in the same place where the spare tractors are stored. The reality is that, most of the time, fields get overused when they are needed, and a constant outflow of cash follows in an attempt to bring back the worn areas.

Given the need for higher-intensity use and the impossibility of obtaining more area to absorb that use, the solution to increased productivity in this case may be a short-term cost increase with a long-term payoff: improving the fields themselves to take greater wear. One obvious though prohibitively expensive answer is artificial turf. A more likely alternative is the retention of the natural turf, improved by methods such as the sand slit process, originated in Scotland. With this system, major tile lines are cut through the field for increased drainage (lack of drainage seems to be the most prevalent cause of field decline), and subordinate drainage feeders are cut as dense patterns of narrow, sand-filled trenches. Grass grows over these, covering them completely. The addition of irrigation completes a moderate-cost solution. Fields refurbished with this technique have been quickly restored to service. Their use capacity has also been increased dramatically.

Major universities spend very heavily to maintain the best possible playing fields for their revenue-producing sports. Their techniques are the targets to learn about. Contacting and visiting the management and maintenance personnel at those schools is a good investment.

Assuming fields themselves are improved to bear increased traffic, the installation of lighting can extend play hours further. Particularly in climates where heat in the summer severely reduces use potential, lighting should be a first priority. The use of utility-grade wooden poles can provide a significant saving over more elaborate metal structures as long as design and placement are carefully considered.

The designer's ability to wrest increased productivity from resources available for recreation has to stem from complete knowledge of the resources—which is only good planning practice anyway. The greater the pressure on the resources, the more important the inventory of data becomes. Information on numbers and kinds of users and on type, character, and quantity of programmed activities, as well as the data about the ecosystems supporting the activities, deserves constant attention and improvement. It is the mine out of which the recreational ore comes. Every effort to make the inventory more complete and constantly accessible will pay off. The use of computer files to compile and keep data, and to compare data to reach new conclusions, is not an option but a requirement for owners and designers. It is neither difficult nor expensive. The real expense results from not using every possible means to discover and make use of every possible resource.

7D. Use of Appropriate Structural Materials

Often, seemingly high initial construction costs can amortize themselves many times over through savings in later maintenance. Conversely, skimping on the construction budget can increase eventual costs by creating maintenance problems. Considering that the money for both construction and maintenance comes out of the same pocket, such long-run viewing justifies many design suggestions that might appear somewhat extravagant at first glance.

This is quite often true in the selection of building materials. For the base plane, an appropriate selection evolves from an analysis of the type and extent of proposed usage. For instance, the designer should be able to determine what the traffic patterns will be. If the pathways people are going to use are not surfaced, they will turn to mud after every rainstorm; they will look messy and be unusable. There are two alternatives. Occasionally disguise the pig wallow with gravel pitched from a wheelbarrow, charting the expense on the maintenance ledger, and watching it exceed the initial costs of a permanent pavement. Alternately, put down the permanent pavement to begin with and busy your maintenance personnel elsewhere.

Putting the paving where the path is seems a perfectly straightforward solution. It is. However, caution is offered at this point against complacency after the pavement is poured. User groups change in age, social composition, interests, and all the other characteristics reviewed in previous chapters. Cursing the customers as inconsiderate blockheads doesn't help maintain a newly worn path (someplace other than the first pour). What we are all frequently guilty of not considering is the meaning of the change. The old path may no longer serve a need. The choices are still pave or suffer, but before picking one, the overall design of an area should always be considered. A path may disappear by attrition if an improperly sited use is moved (Figure 4-30 on p. 68).

Assuming the overall plan has been reviewed and the nature of use has been predicted for the foreseeable future, the next step is to select a material that can withstand the rigors of the expected activity. A material is happily matched to its use if it passes the following tests:

Durability. Will it stand up under the anticipated pounding?

Appearance. Is it visually compatible with nearby elements?

Availability. It is economically foolish to haul material from distant sources if comparable material is locally handy.

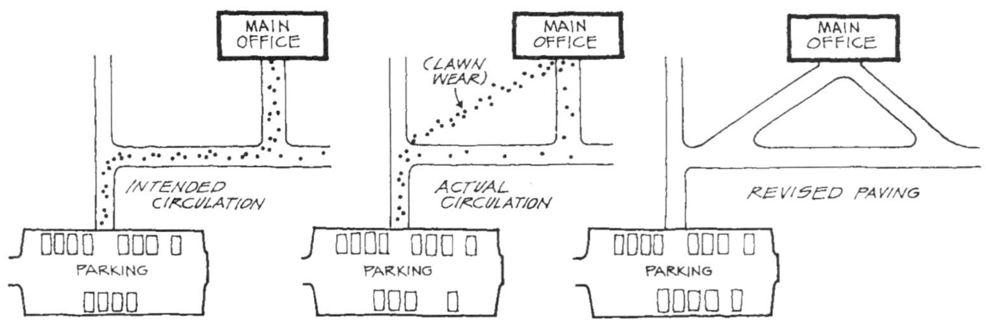

Figure 4-30
Paths in the right places.

Tactile qualities. Its feel is especially important when the material will come in contact with the skin, as in the case of sitting and playground surfaces.

Climatic adaptability. Will the material remain stable under such rigors as freezing, thawing, and intense sunlight?

Drainability. Does it allow rainwater to percolate through or run off rapidly and render the area usable after storms?

There are a host of materials to choose from: concrete, brick, cobbles, asphalt, wood blocks, gravel, sand, grass, Astroturf, and others, with many varieties of each. When matching a material with a use, the above criteria should be applied along with whatever priorities are suggested by the activity. For example, in a playground you might consider concrete for a peripheral walkway, since durability is the most essential criterion, and place a soft granular material under the equipment where softness and quick drainage are of paramount concern. (See Figure 4-31.) Or, in a neighborhood picnic ground, you might decide on grass for the unpatterned playfield, gravel on the paths connecting each picnic unit, and a hard surface like concrete under each picnic table and fireplace station.

Figure 4-31
Surface treatments related to use: soft and drainable under the play piece; durable where constant foot traffic is expected.

Suppose a hard surface has been the choice and the options of locally available materials (a *very* important criterion) have been narrowed down to brick and blacktop (the dilemma mentioned in chapter 2). The designer selecting the best-looking material need not take lumps from a critical client for being extravagant. Paraphrasing an old joke, the benefits of purchasing material today because it's cheap, even though it's ugly, have to be weighed against the vanished cost differential after 10 years' maintenance—when it's *still* ugly.

In fact, faced with cost concerns but having an equal concern for aesthetic standards, a Midwestern university found that brick paving for pedestrian circulation could be laid fairly inexpensively by laying it dry (no mortar joints) on a modest 4-inch-thick concrete slab. Compared only a few years later to asphalt paving installed for a pedestrian walkway at the same time, the brick not only looked great but weathered well (asphalt didn't), wore well (asphalt didn't), didn't ravel at the edges (asphalt did), and didn't show patch marks after utility lines had to be sliced through it (asphalt *really* did). The bottom line showed the dry-laid brick as the winner in both the aesthetic and cost columns.

The challenge in making decisions about materials is the same as in making decisions affecting people. Keep learning more about them and beware of facile assumptions. Critical observation is critically important. Look at nature's treatment of materials. The reason native materials are native is that they live well in their locale. Unlike concrete, which is a composite of not necessarily local materials, stone in its native setting may be an extremely good choice aesthetically and economically because it belongs there—it doesn't self-destruct in a winter that long since gave up doing stress tests on it. Granite curbs, though initially expensive, continue to grace the edges of many East Coast roads because they pay off as they outlast most of their concrete competitors.

However, concrete in the form of precast paving units, popular in Europe for more than a half century and almost as long in the United States, has some distinct advantages in flexibility and cost for recreational paving uses. This type of concrete paving unit is created at extremely high density and consequently has a high resistance to environmental damage. Laid on a compacted granular base, the paving can include elegant patterning in muted colors and can be removed for underground repairs, retained indefinitely, and replaced with no visible sign of disturbance. Bill Schneider, president of LPS Pavement Company and a nationally recognized expert in precast paving systems, emphasizes in every case the vital importance of a well-placed base course. With that accomplished, he points out the extraordinarily good record of low maintenance and high durability of paver systems for anything from airport runways to bikeways.

SUSTAINABILITY SIDELIGHT 1

What's the best paved surface?

In any comparison of paving types for hard surfaces, the essence of the question is long-term use of the area. The less hard the surface, the greater the repeated replacement. From the then-elegant engineering of the Roman roads, which were simply crushed stone laid very systematically with tiny stones securing the larger bits on the surface, paving has in some way been an aggregate (usually stone chips) held by a binder. Asphalt as a binder holds the aggregate but remains a liquid, even though highly viscous. Cement as a paste holds the aggregate and hardens into a rock (concrete). Because it does truly harden, concrete paving or concrete unit pavers offer the best solution. Even if the area is small (picnic table sites, for example) it's worth the design effort to group them on a single surface and *use pavers for their design flexibility.*

SUSTAINABILITY SIDELIGHT 2

Water management equals free land.

Hard surfaces are necessarily part of new construction (roofs, parking, roads, drives, walks). Even our parks, with play fields and open space, have huge hard-surface areas for the entrance drives, roads, roofs of structures, and parking areas. The play fields frequently have artificial turf and extremely smooth earth surfaces (like ball diamonds) that are almost as impervious as asphalt. The runoff escapes the site and turns into part of downstream floods.

Ideally, water that falls on the site stays on the site. Roof "gardens" do help detention by slowing runoff, Retention ponds are basins sculpted into grassy areas and runoff water is channeled into them, to be retained and allowed to evaporate or seep slowly into the earth. Detention ponds detain water by slowing the flow of water out of the pond through deliberately small pipes or channels. These reduce flow satisfactorily, though at the cost of loss of large and useful areas of open space.

Parking and paved areas for circulation, open pavilions, picnic table areas and other group use areas can be paved with pervious concrete paving units which does radically reduce runoff from those areas (down to 25% and generally less). In addition, the water can be stored in the base of the paving, eliminating the need for separate retention basins and reducing much of the sewer piping necessary to the typical hard-surfaced drainage systems. Fewer basin commitments? More land available for more park activities.

7E. Comparisons of Pervious Pavement Types

Pavement that allows the percolation of rainwater is generally based on three types: asphalt, concrete, and precast concrete pavers. Both the asphalt and concrete, when used in this way, must have spaces between their particles of aggregate. Due to the openings, they are both structurally weaker than the unitized concrete blocks, which are extremely dense and hard. Depending on the traffic load, asphalt may have a 2- to 3-year span between repairs. The concrete can have a 5- to 10-year span before repair or replacement is needed. The concrete block paving system has a projected life of 50 to 60 years. Costs of the paving types varies from about $1 per square foot for asphalt, $3 to $5 for concrete, and $6 to $10 for the unitized pavers.

The most successful and maintenance-free systems are generally known as *unitized permeable concrete pavements*. They have become an especially valuable material choice in the last two decades because of their adaptability for providing a pervious surface while still providing a structurally sound and attractive surface for all types of traffic loading. Unlike the other pavement types, the damage done to part of the paving by having to cut through for a utility line or repair to subgrade services does not leave an ugly repair scar. The pavers are simply lifted from the work area, the work is done, and the paving is restored with the original units—and no trace shows. Overall, the maintenance cost of the unitized pavement is far lower than other pavement types.

Regarding the issue of stormwater management, the unitized pavements have deep bases (12–18 inches or more), composed of very coarse rock with an average of 40% void space for water retention. A parking lot of about one acre in size with a 150-car capacity can easily hold a 2.5 inch rain in 3 hours or over a full day. The retention pond that would be required to hold an equivalent amount would be an area adequate for a tennis court or a dozen more cars. That's free land not otherwise available to the park.

In hot and dry areas, porous pavements of all kinds have great value for their retention capability. The retained water is available for storage, irrigation, or gray-water applications. Other benefits include the reduction in surface evaporation loss, since the percolation is extremely rapid. Some benefit is also derived on the surface from the temperature modification created by the subgrade water mass itself.

Whether for large or small areas, unitized porous pavers used to drain and retain water are definitely worth evaluating. Significant water retention is accomplished and surface-to-sewer drainage is hugely reduced. Repair or replacement? Seldom necessary. Cost savings? Generally enough to beat the continuing cost of the replaced grass, mulch, stone chips, or whatever. And it looks great.

In 2003, faced with a long list of tough decisions to guide a huge investment, the Morton Arboretum in Lisle, Illinois, had to find the right way to redevelop their main parking lot. It needed to accommodate 500 cars and buses, up from the original 160. The new project neighbored with Meadow Lake, a scenic gem which not only couldn't be impacted by the work but actually had to have its water quality upgraded. The choice to use the unitized permeable concrete pavement system resulted in 200,000 square feet of attractive, durable surface. The pavement covers a highly successful filtering system through which all the rainwater passes, carrying non-point-source pollutants to the filtering system and cleaning the water before it enters the lake and nearby river. Based on its "green" value and demonstration of potential, the state offered a grant of $1.2 million in support of the project, which is regularly used for workshops to help other agencies and developers learn about the system. Maintenance for the elapsed decade had basically been an occasional vacuuming to remove surface litter.

Grass and Alternatives on the Base Plane. Grass poses particular problems as a base-plane material. It does well only where it is left untrampled or can be given a chance to recover from the beating it takes from traffic. Thus, if movement is anticipated in a constant line from A to B or is concentrated in a particular area, the immediate vicinity of a picnic table for example, grass quickly gives way to mud and weeds. One solution to this problem is to move the tables periodically. If this is impractical, base-plane materials other than grass are mandatory. These might range from pavement to wood chips or other organic mulch. The mulch requires periodic replacement, but, unlike grass, it does not demand spraying, fertilizing, and mowing. On the other hand, grass might be quite appropriate where use is intermittent or movement from A to B follows several courses with no single pattern predominating.

If mulch begins to become a major material need, some creative effort may unearth some inexpensive sources. Many utility companies clear brush and trees from their rights-of-way and chip the material to ease disposal. It may be free for the hauling. Some wood-processing companies may do the same. Other types of mulch should be considered, such as shell mulches from nut-processing companies or ground-up cob mulches from corn-products companies. What else is going to waste locally? If bush and tree removal is a significant item on a park agency's work list, the purchase of a chipper-mulcher may be an excellent investment.

The base plane offers many other possibilities for creative use of structural materials. (See Figure 4-32.) Imagination and willingness to seize an opportunity can help change a truckload of jettisoned brick from junk to be dumped to paving material for a craft court. A street of granite-block stone removed for building construction becomes a mine of surface material for paving large or small areas creatively. The success of any paved area is dependent on the success of the base beneath it, just as with the unitized concrete paver installations. Broken concrete sidewalks become riprap for stabilizing erosion-prone slopes while vines get established in the interstices. (The teeth-clenching aspect of this creativity is the unfortunately prevalent surprise factor. Nothing seems to become available quite in the form or at the time that might be just right.)

Figure 4-32
Can a special need be satisfied with an apparently "unspecial" material?

The base plane is also dirt (if you don't need it) or earth (if you need more). The sight of trucks full of topsoil driving by on their way to a spoil site further away should inspire you to invent some use for that soil right at home. If a place can be found for it, the soil might be free since the hauler has to pay more to take it a greater distance. Consider a "dune" concept: The mound might screen parking or service areas or provide a wind buffer adjacent to a senior citizens' area. As in the trees on wheels scheme, some useful purpose could be served in the right holding location, even though the soil changes shape for a time until it is permanently used (Figure 4-33). Temporary seeding can stabilize the surface. The increased cost of the material later (if you can find it at all) offers a strong inducement to this kind of flexibility.

Figure 4-33
A dune for topsoil storage?

Figure 4-34
Plants can form spaces.

Figure 4-35
Plants can direct circulation.

Figure 4-36
Plants can provide detail interest.

Figure 4-37
Plants can deter wind.

We will look next at the choice of vertical-plane materials. Here, weathering is the primary culprit creating maintenance problems for fences, walls, and building facades. To minimize long-term expense, it is therefore wise to select materials that need little attention in order to withstand the bombardment of wind, rain, and sunlight. This might mean choosing rust-resistant metals for fittings and fastenings, and stone, brick, and concrete in lieu of wood.

It is also important to cautiously assess the after-effects of material selections and to make corrective changes if necessary. One agency adopted a system of elegant and rugged metal signs made of cast aluminum letters fastened to painted steel faceplates set in aluminum bar frames. The painted panels almost immediately bloomed with rust around the edges. After removing, repainting, and replacing the signs (several times), the agency realized that the fabrication and painting were not at fault. The contact of the steel panels and aluminum frames produced an electrolysis of the two metals that no paint could correct. After aluminum sheets were substituted for the steel faceplates, no more problems occurred.

Where wood is used, tinted stains provide a finish that needs only an occasional refreshing, whereas paint demands recoating at more frequent intervals. In addition, the knocking about that occurs in any public place guarantees that painted surfaces will be chipped and flaking long before they are recoated. However, under the same punishment, stained planes will slowly fade, thereby preserving a reasonable appearance until they can be restained.

Although expensive, redwood and some of the other tropical woods offer a nearly ideal material, for they not only require no finishing but are actually enhanced by weathering; wind, rain, and sun turn their unfinished surface to a soft gray after a few years of exposure.

Technology of plastic materials has progressed in the use of recycled resources to the point of becoming a structurally desirable alternative to the use of wood in appropriate situations. Recycled plastic "timbers" are available in standard lumber sizes and have the structural strength as well as the handling and working characteristics of timber. They can be cut, shaped, and fastened with traditional materials and techniques and offer excellent weather resistance.

7F. Use of Appropriate Plant Materials

Long-term maintenance expenses also can be minimized by the selection of proper plant materials at the time of design decision. Good judgment here results from a recognition of both similarities and differences between plants and other materials. Like lumber, stone, concrete, and other inanimate materials, plants can define static and directional spaces; provide human-scale detail; screen wind, sun, and less desirable views; abate noise; control erosion; and channel circulation. Accordingly, plants, like other materials, must be of such size, shape, and staying power as will suit the job they are given. (See figures 4-34 through 4-41.)

But plants have unique credentials that set them apart from inanimate materials. Their seasonal changes lend year-round variety that cannot be imitated by nonliving objects that remain first and always as they were set in place. The same plant can be green of leaf in the summer, gold in the fall, laden with berries, branching intricacies, and twig color in the winter, and a kaleidoscope of bloom in the spring.

Further distinguishing plants from other materials is the fact that, as living things, they must be maintained similarly to the way we treat our bodies, which is quite different from the manner in which we keep up our houses. Each species requires specific environmental conditions for survival. Species can be categorized according to *soil needs* (heavy, light, acid, alkaline), *moisture needs* (constant drinker, seldom thirsty), *exposure* (tolerance to sun, shade, wind), and *hardiness* (ability to withstand extreme temperatures). If a plant requires one condition, it will suffer under the opposite. Species that thrive in a swamp will perish if placed in the desert.

In addition, plant species can be cataloged as to *life span*, *susceptibility to certain diseases and pests*, and *ability to survive surface compaction and fill*. If these clues about the plant's lifestyle are disregarded, countless hours must be spent assisting the plant in its fight against its environmental enemies. And, even with such help, the species may not make it.

Therefore, why not try to avoid all but the minimum amount of fertilizing, watering, cultivating, spraying, and wrapping with burlap to ward off the winter winds by planting only those species which appreciate the conditions found on the site. Let nature do most of the work for you. You can't beat the prices. And this is an excellent example of the most basic definition of sustainability.

To take advantage of nature's maintenance service, a useful rule of thumb might be: If the site has a great deal of plant life on it, select for the new inclusions only those species which normally fit into the *community* of plants in evidence. Under natural conditions, plants survive by complementing each other; for example, the sun-loving reach above the rest, shadowing the shade-loving. Those who do not contribute to this cooperative arrangement are soon eliminated. To thrust a foreigner among an existing community is to maximize the risk that it will not survive. This rule has its aesthetic benefits as well, for putting like among like helps maintain the prevailing effect.

In the city, where development interferes with the growth of natural communities, plants should still be selected to match the existing (albeit artificial) conditions. In addition to criteria already stated, such considerations as *tolerance to air pollution*, *salt spray* from icy streets, and *pavement reflection* must direct the selection of city plants. Pavement presents a double problem. Besides throwing sun glare, it prohibits rainwater from reaching roots. It is therefore a guaranteed loss to pave right up to tree trunks. Paving should be at least a 6-foot radius away from the trunk. The installation expense is well justified by future savings in maintenance or replacement costs.

The cost of a free gift deserves a caution in this regard. A tree donated by an owner, even though a dandy and desirable specimen, has to be looked at in terms of the cost and formidable hazards of a move: First considerations should be the "from–to" factors. Can you get *to* it with the necessary equipment? Can you move it *from* its existing location *to* its new home without collapsing the owner's basement wall, dropping utility lines, moving parking meters (easy to overlook), or, worse, having to cut another tree back to a hat rack to make space for this one? Even if *you* can accomplish the necessary physical changes, can the tree? It is being moved from an unrestricted soil surface at its source location to a new environment that may challenge it with a hostile soil, a severe exposure, a skimpy water table, and an affectionate audience waiting to love its root surface to death as they play under its branches. Once in place, the care necessary to nurse it along may be the greatest expense of all. Some examples of regional guides to the selection of appropriate landscape plants are included in

Figure 4-38
Plants can supply shade.

Figure 4-39
Plants can buffer odors.

Figure 4-40
Plants can suffocate noise.

Figure 4-41
Plants can retard erosion.

Figure 4-42
Plants have predictable forms.

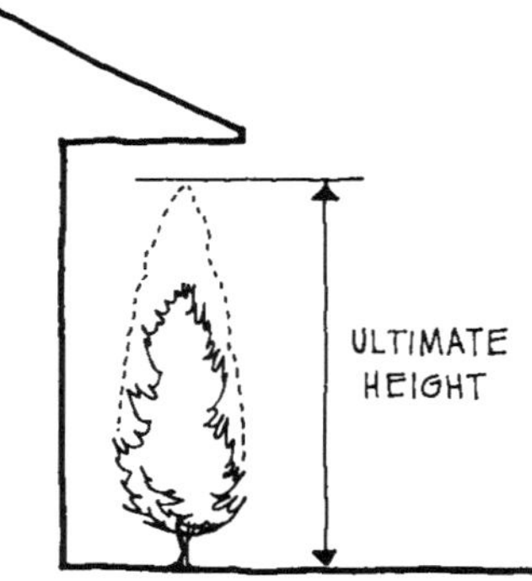

Figure 4-43
Plants have predictable growth rates and sizes.

Figure 4-44
Without undergrowth to take over when the older generation of trees dies, the life of this picnic site is limited.

Figure 4-45
Linear nurseries for beauty with easy service access.

Appendix 4. Considering the variety of conditions and locations involved, turn first to the web to guide your considerations for your particular situation.

To minimize maintenance chores, the most obvious criteria should also order the selection of plant material. That is, first of all, *plants grow*. Plants should be chosen that have growth habits naturally fitting the circumstances into which they will be placed. (See Figure 4-42.) This will eliminate the extensive pruning required to hold a plant to atypical form. In days gone by, horticultural fanciers got an immeasurable charge out of plants shaped into globes, cones, and various forms of animal life. Taste aside, perhaps they could afford the grooming effort. With today's burgeoning costs, however, it should be recommended that if one wants to spot a chicken effigy in the park, one should consult a taxidermist rather than a horticulturist.

Along with habit, rate of growth and ultimate size should also be considered. The thinking here cuts two ways. If the selection is slow growing, do you have enough patience to wait for it to fill its appointed place? Yet, if it grows too rapidly, how long before it devours the utility lines or lifts the roof overhang from its joists? To avoid having to wait years for appreciable effect as in the former case, or butcher major limbs as will be required in the latter, it is wise to select a species that will grow at a pace suited to the circumstances—slow if the spot where it is planted is a cramped corner, fast if it must throw up quick screening—and fit its allotted space upon maturity. (See Figure 4-43.)

Plants have limited life spans. Replacement to maintain the intended character or effect must be continually encouraged. An administrative and maintenance policy can be implemented whereby picnic and other areas of intensive traffic are retired periodically to enable underbrush to establish itself. As illustrated in Figure 4-44, a defeating policy would be to wipe out the scrub because "it looks neater." The young material, however disheveled in its infant state, is the best replacement stock and is already in place. Reaching maturity while the taller members decline through old age, the young stock will be ready to make a substantial contribution when the oldsters expire.

A park agency can use this kind of natural nursery approach to great advantage if new growth can be integrated with older materials in casual settings. Other opportunities for developing plant material resources on agency land should be considered also. Since plant replacements and additions are constantly needed, areas suitable for holding, growing, and transplanting trees should be identified for use but not necessarily developed as typical row-by-row nurseries. Recalling the multiple-use concept urged earlier, consider nurseries also serving as windbreaks, view screens, use separators, educational arboreta, and sound buffers. Comparatively narrow right-of-way strips can be planted in softly linear, changing patterns, making the parks' drives and trails delightfully scenic while offering easy access for placing and replacing the trees (Figure 4-45). Once suitable locations with flexible boundaries have been identified, a nursery sellout, a tax auction, a donation, or a discovery can be seized as an opportunity and the trees obtained stored with immediate *and* long-term advantages.

Recognize also the opportunities inherent in relationships with research agencies like timber corporations, major nursery companies, and state or federal as well as university departments, all of which may be doing growth studies of large-scale plant materials. Trees developed in these agencies' programs may be intensively used for 5-

to 10-year study cycles and then, having passed their test target limits, be disposable. Rather than letting them be cleared away, another agency with an opportune place and purpose may be able to have them purely for the taking. Being aware of available resources is again a key to capitalizing on this kind of opportunity. In fact, an offer of one resource (land) in exchange for another (the plantings) could conceivably lead to the research project being developed on the target site where the trees are ultimately needed. Everybody wins; research results are obtained and the agency gets plant materials without needing even to transplant them.

At this point, it would be appropriate to view all the planting being done in any park as a *transplanting experience* (unless you're planting acorns for that oak allée).

All the materials planted in a park will be delivered in a visible/invisible condition. The plant's visible, above-ground parts will include, probably, a trunk, and definitely branches. The invisible parts will be its roots, whether delivered bare-root (as though summarily yanked out of the ground), in a container (a pot, bucket, tub) which will contain the roots and also the soil in which they have been held, or in a ball (consisting of the roots and the very ground in which the plant was growing from its start), dug up, wrapped with a sturdy fabric (or boxed) so that the roots remain intact for the move.

The ultimate success of the plant depends on three things: its new soil home, its long-term source of moisture, and ratio of maintenance it needs to the maintenance it receives. Consider these items in reverse order:

The maintenance it needs, once established, should be . . . zero. Why? The plant, chosen with sustainability carefully considered, would be native to or fully adaptable to the new location. The local soil's nutritional value and the local precipitation should be similar to the plant's origin. Then with a period of a year or so of supplemental watering and fertilizing as appropriate, the plant would be growing into its new home. Admittedly this is the ideal, but in fact it's not that hard to accomplish in an open park site.

But many park sites, and more to be coming, are not simple open fields. Parks integrated into cities and suburban areas have extensive areas of paving. Even the open areas are intensely active with foot traffic and maintenance vehicles pounding the surfaces. The resulting compaction of the soil in which the plants are living creates a soil in which the plants are dying.

This problem, with the evidence showing as skeletal remains of sickly trees along city streets, parkways, and in the plazas of city parks, became such a frustration to James R. Urban (an incredibly appropriate name), landscape architect in Maryland, that it has become a crusade for him. Through decades of field observations, extensive photography, and comprehensive testing procedures, he has developed a well-documented process for identifying and correcting soil and root problems. He has also documented and published successful tree planting procedures and specifications. His book, *Up by Roots*, is an award-winning guide to insuring success in urban plantings where the trees are vitally important to the success of the design and the construction of the design is vitally important to the trees. At the core of this concern is not only the importance of the appearance of the resulting parks, but, more important, the sustainable blessing of minimizing long-term maintenance through a good initial investment in proper installation.

Figure 4-46
Plants can provide a wind wedge for energy conservation.

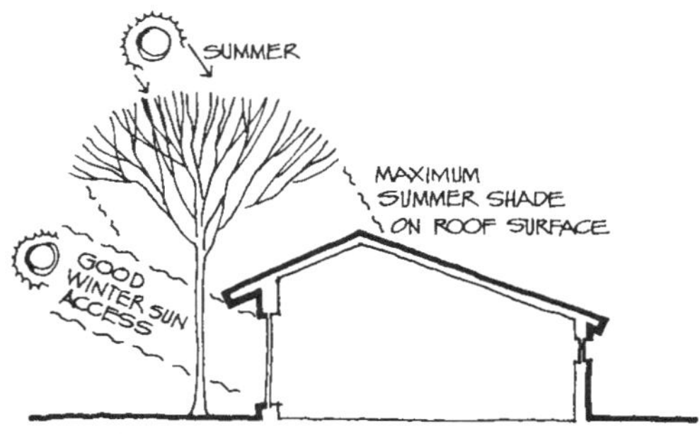

Figure 4-47
Shade trees, if properly selected and placed, are solar thermostats through the seasons.

Figure 4-48
Smooth slope edges (below) are both practical and visually appealing.

Figure 4-49
Paving at the base of a wall accommodates one wheel of the mower, thereby leaving a cleanly cut edge.

7G. Energy Conservation

As mentioned in the discussion of orientation to natural forces, plants offer unique possibilities for energy conservation through control of weather and natural forces. Combining some of the plant abilities previously mentioned, a structure can be buffered from severe winds by mass plantings of dense evergreens, particularly if these are selected by size and species to offer a wedge to the wind (Figure 4-46). Shade is also a vital commodity in an air-conditioned. world. Trees selected to provide energy-penurious shade must have the growth potential to reach up and out over the structure. They also have to be accommodated in a location that allows them to take advantage of that potential. A shade tree intended to serve a structure in this capacity may seem to violate conventional design patterns by hugging the building and by being limbed high (by nature or by maintenance) as in Figure 4-47. This confrontation of good design with preconception brings us back to the oft-repeated key word *why*. Preconceptions can be dangerous obstructions to a good answer.

7H. Attention to Details

If base-plane materials appropriate to use, vertical-plane materials that will weather well, and plant materials appropriate to the location are specified at the time of design and if materials are placed so as to take every advantage of energy conservation potential, the maintenance budget should breathe a sigh of relief. These are basic matters of concern to which can be added many minor considerations for further savings.

For instance, slopes must be given special attention because they are the surfaces most likely to erode. Intensive traffic should be kept away from steep banks because constant tromping will obliterate the cover that holds the soil in place. Grassed slopes should be rounded at the tops and bottoms, as if stroked with a butter knife, rather than shaped with abrupt edges, as if hacked out with an axe. As indicated in Figure 4-48, such rounding provides for both mowing ease and a smooth visual transition.

One more note on the subject of mowing: Figure 4-49 shows the advantages of paved edging at the base of vertical planes. The mower can cut clean. There is no need to go back with hand clippers to get at the tufts that would have been left by the mower at the base of the wall had the edging not been in place. Granted, a string trimmer would be an easier alternative than hand clippers, but the target is to *eliminate the entire step*.

Raising planting surfaces (or depressing walks) can deter people from short-cutting through and stomping on the plants. Similarly, depressing sand areas in playgrounds helps to keep the grit in its proper place. Since sand is a fluid material, it will eventually assume a plane parallel to that of the horizon no matter how it is heaped originally. Therefore, to avoid a slumping overflow, the edges of the container in which the sand is placed should be level.

In the category of many unhappy returns, trees planted by a park agency in sidewalk openings adjacent to a street curb were being carefully protected by an epoxy-bound aggregate surface installed over the entire opening (about 4 feet square). The project was well conceived: The aggregate, about the size of pea gravel, makes an easy and attractive walking surface for pedestrians; its porosity is excellent, allowing good air and water circulation to the tree's roots; and the epoxy binder itself has been found to have a good survival capacity in severe weather conditions. The quality of work was good; installation involved making a neat ring around each tree, about an inch away from the trunk, with a reusable form. However, if the trees have any success at all, they will grow smack into the edges of their holes in two years maximum. Obviously the holes will have to be enlarged. But to what extent? The careful work is wasted. Since no expansion cuts or lines were provided to allow enlargement in precise steps, a jagged mess will have to be made of a good job. Growth anticipation should obviously be a factor in design of any covers for tree openings, including steel grates or cast materials of any kind. Ask yourself: How can they be made to grow with the tree?

In response to the continuing popularity of bike usage, a number of new types of bike racks have appeared, although many varieties are still based on the traditional bar frames—that is, they have some kind of separator placed vertically at about 6-inch intervals (wood slats, steel bars, concrete slots, or aluminum tubes). The theory of the design is that the small module allows flexibility; any number of bikes can be placed anywhere on the rack. At one university campus, however, comparisons of the 6-inch, all-purpose racks to more apparently inflexible structures with openings 2 feet on center showed that given the choice, people spread their bikes *farther* apart in the 6-inch interval racks. Intervals of 30 inches, 3 feet, or more meant that actual capacity dropped by as much as 50%. The racks with 2-foot spaces, however, were used at full capacity.

Garden hoses typically come in 50-foot lengths. Two can be fastened together but when a hose is longer than 100 feet, the friction in the hose itself reduces the outflowing water to a feeble trickle. On a property without a built-in sprinkler system, if the only available water spigots are 200 or more feet apart, there are places that only nature can water.

These are but a few examples of the kinds of details that should be investigated in any design solution. Small in execution, they loom large in significance, for attention to them can prevent major maintenance problems.

Principle 8: Provide for Ease of Supervision

Freedom. It has been suggested that to be unencumbered by overbearing authority is a need common to all. This immediately appears to conflict with an administrative requirement, the need to have people use an area as intended. Obviously, use must be supervised to some extent. And every public place has its share of legitimate *don'ts*. Don't tramp on the flower bed because you will make mud. Don't enter the complex at arbitrary points because this will make it impossible to collect fees. Don't run through the archery range since you may end up with an arrow in you.

But *don't* is a culprit word. It nags. It looms as a challenge. In many cases, it promotes the misconduct it seeks to put down. What is your impulse on spying a *Keep Off the Grass* sign?

To gain the necessary control, yet retain for the user a sense of freedom, landscape architects attempt to replace *don'ts* with *do's* by organizing use areas and circulation routes in a manner that will make it appear reasonable to use the facility as the designer intends and the administrator desires. This is the easiest way to facilitate supervision—let design layout guide visitors into a use pattern that they will find agreeable.

Matters of Concern

8A. Balance of Use Freedom and Control

The degree to which control can be exercised should be addressed early in the design process. It will be found that there are circumstances in which any attempt to control movement or use is a waste of effort, for the directive will be ignored. The simplest example is the situation in which movement patterns cannot be predicted. Where will the user penetrate the large field in order to reach the other side? Unless an obvious purpose such as safety is served by stepping off at a set point, it will be wherever whim suggests. In this situation, it is folly to demand a single entry point, for the regulation will seem unreasonable. Design should allow whim to take its course, because it will anyway.

There are also instances where use not only cannot, but should not, be regulated. Consider a free play area. To the users, it appears to be an arena for doing their own thing. What becomes their attitude, however, when they come across the signboard that spells out what their thing shall be—and also dictates what it shall not be? Shouldn't there be places not only in every system but also in every park where one can express individuality, act freely, exercise discovery mechanisms?

Actually, all we propose is that parks provide the opportunities formerly available on the late, lamented corner lot. Here was gained a release not possible on sites where there is only one way of doing things. Here was freedom of prescribed choice—truly free space. The lot is no longer with us. But is that reason to do without its rewards? Cannot sites or portions of sites be let simply to sit there, beckoning the user to wander, dig a hole, build a shack, fly a kite, or bake a potato? It's the user's choice. It's the user's recreational appetite that needs to be satisfied. It's the user's need that the system proposes to serve.

Who can predict every urge? The recreationist conducting demand studies and the behaviorist spinning theories admit to not even being close. As useful as the information provided by these specialists might be, it will never be absolute. How then do you give the unpredictable its due, if not by simply providing a place and letting it happen? Where freedom of movement and intent is deemed desirable, little design elaboration is necessary except in some cases to ensure maximum use flexibility in the structures and in the organization of the park units.

This is not to suggest that use discipline should never be required, but to stress that restrictions (or lack of them) be for a purpose. Where it is apparent to all, especially users, that reasonable purposes are being served, conflict between the controlled and the controller is less likely to arise, and the chances of the directive's being followed are maximized.

8B. Circulation

In any public facility that serves great numbers of people, movement is an issue of primary concern. If people can get where they want to go readily and without interfering with other activities, a feeling of peace permeates the site. The charge is therefore before the landscape architect: Anticipate flows. Eliminate obstacles and confusion. Provide unobstructed, well-defined, and logical routes.

Obstacles may be natural barriers such as boulders or steep slopes, but they are more likely to be use areas. It is by setting up good relationships among use areas that the designer takes the first step in establishing an efficient circulation system. For instance, as diagrammed in Figure 4-50, if the parking lot cuts off the picnic area from the swimming beach, it is a cinch that the picnickers will stream through the lot on their way to the water, holding up traffic and endangering themselves, no matter how many signs, railings, and other don'ts suggest otherwise. But if, as shown in Figure 4-50, the parking lot is located peripherally so that the picnic grounds abut the beach, the most direct route will still be followed, but it will be a safe one.

Similarly, how can you expect kids to keep the noise down in the play space shown in Figure 4-51A just because it happens to be next to a quiet zone? And it is just as unreasonable to expect that they will tiptoe through the passive area on their way to the ice cream wagon. Let's say that a natural barrier of earth and trees is nearby as indicated in Figure 4-51B. Use it as a buffer. Assign the play space to one side, place the quiet zone on the other, and locate the refreshment stand where access to it from both sides will be unobstructed.

In both examples, the successful relationship patterns suggest *dos*. The logic of the organization itself tends to supervise the movement, thereby easing the way for the actual linkage of the use areas by roads and walkways. It becomes easy to understand how these routes link the use areas if we divide them into three types: *collector* arteries, which connect all major use areas; *secondary* arteries, which lead from collectors to connect related spots within a use area; and *minor* arteries, which proceed from secondaries to the least-visited facilities.

A designer who understands the collector-secondary-minor premise well can devise access routes to a host of activity units in a manner that simplifies a traveler's decisions and minimizes confusion along the way. First, the designer can eliminate intersections and the accompanying slowdowns from a heavily used collector, by clustering many activity units about a single secondary rather than stringing them out along the main road, where each needs its own separate driveway. While all visitors will be using the collector to enter, leave, and search for objectives, only those interested in a particular group of activities need be on a secondary. Accordingly, travelers do not have to plow through activity group X in search of group Y, for each has been segregated from the main flow. (Compare Figure 4-52 with Figure 4-53, both on p. 80.)

Confusion is further lessened if each road unit flows smoothly, for people have a tendency to follow consistency in alignment. This minimizes inadvertent movements from collectors to secondaries or minors and the subsequent need to move back in search of the artery upon which the journey can be continued. The principle can be applied just as successfully in reverse. The temptation to move down fire and service roads from which the public is excluded is eliminated when such arteries are made to connect with the public way at right angles or otherwise break the main road's continuity.

The way in which turns are treated can often mean the difference between free sailing and eternal congestion. Turns should never be more acute than 90 degrees because at each intersection a decision is being made and the accompanying hesitation, if coupled with a turn that is difficult to negotiate, could cause a traffic tie-up. In addition, each intersection should be cleared of visual obstacles to give a good view of oncoming travelers.

Classic tie-ups are also common where left-hand turns across traffic are required. Hence, turnoffs should be to the right wherever possible. This is of particular significance for the entering sequence, for it

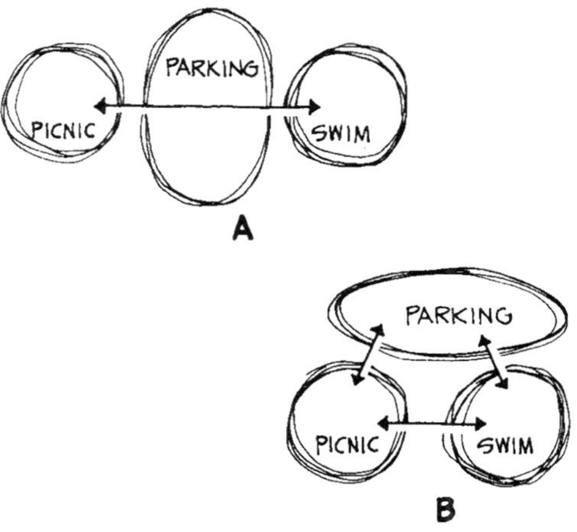

Figure 4-50
Use-area organization can encourage either an undesirable (A) or an agreeable (B) traffic flow.

Figure 4-51
A poor site relationship system (A) can he revamped by taking advantage of an existing site characteristic (B).

is on entering that the traveler, unaccustomed to the layout and searching for objectives, is most confused. A series of left-hand turns, going against the normal right-hand grain that most of us possess and forcing drivers to turn across vehicles speeding from the opposite direction, only adds to that confusion. However, left-hand turns are comparatively inconsequential in the exiting procedure, for by then the traveler has a better feel of the layout and has a single objective, the exit.

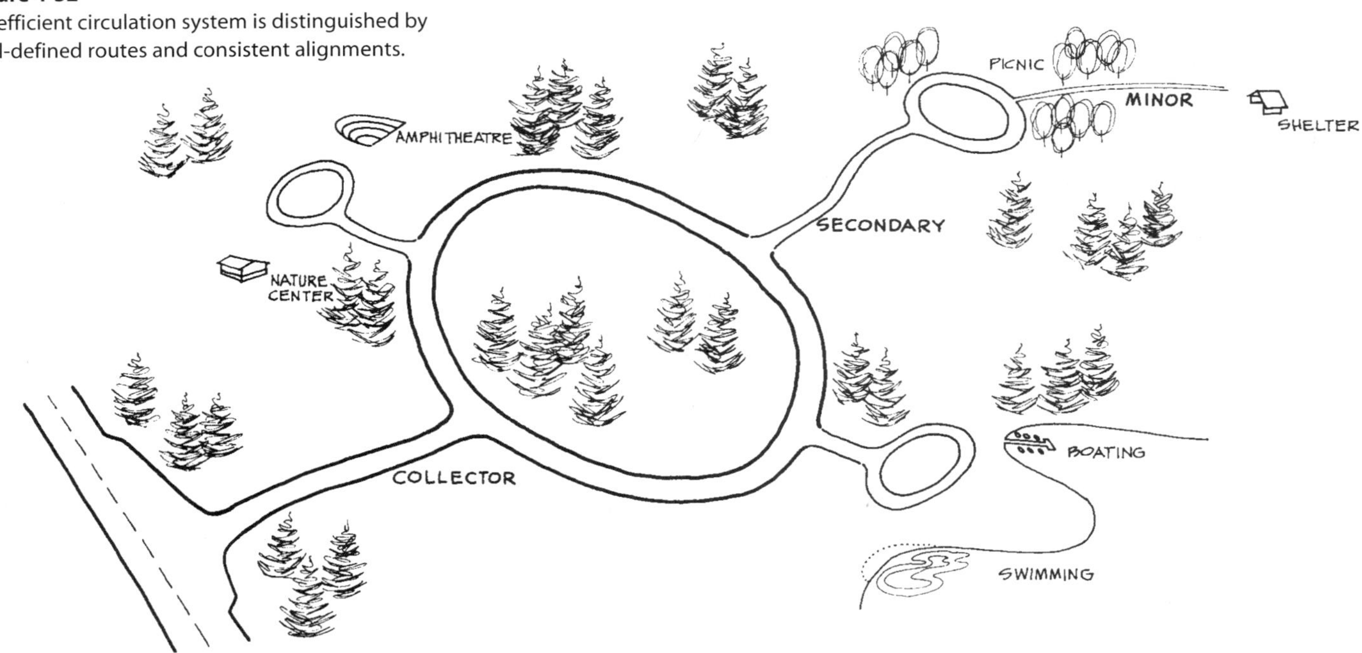

Figure 4-52
An efficient circulation system is distinguished by well-defined routes and consistent alignments.

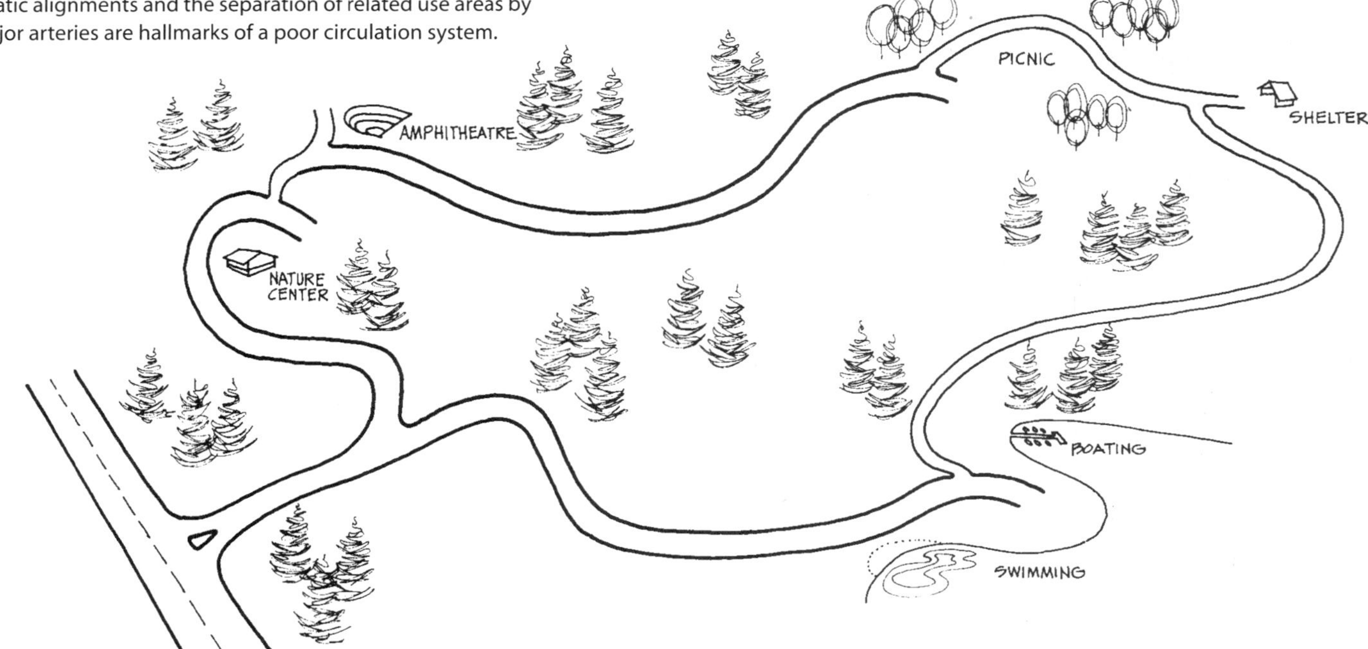

Figure 4-53
Erratic alignments and the separation of related use areas by major arteries are hallmarks of a poor circulation system.

To help the visitor adapt to an unavoidably confusing layout, points of orientation might be inserted. These could be outstanding pieces of architecture (Figure 4-54) or natural features that the scheme would cause to continually pop into view. Points of orientation are handy references that let the visitor know where he or she is at all times: north, south, east, or west; close to, distant from, in front of, or in back of something frequently in view. These references are especially useful in such places as a zoo or fair where there are so many facilities to visit that the possibilities of becoming spun around and lost are high.

What many of these devices and patterns are actually doing is giving information, suggesting to the traveler: Go this way. Go that way. You can find your parked car over there. An abrupt change can also provide information, giving travelers notice that their attention will soon be required. Such changes as dramatic movement from linear to static space, light to dark, human to superhuman scale, smooth pavement to rough, jar the mind, thereby ensuring pedestrian alertness at the approach to a potential danger such as a vehicle crossing. (See Figure 4-55.)

Another kind of change that subconsciously informs is a change in elevation. A slight change such as a curb suggests *keep off*. Unlike the *don't* of a sign, this type of suggestion should breed cooperation. Because it is an effort to mount the curb, the traveler will most likely agree to remain in the depression. Where the requirement is mandatory, the vertical can be increased in height in order to say *keep out* in a stronger voice. The ultimate directive is a vertical elevated above eye level, the barrier in front of the eyes leaving no doubt as to the intended circulation pattern.

Out of intelligent use relationships, a simplified collector-secondary-minor artery system, and such good design details as negotiable turns, right-hand movements upon entering and at other points of vehicular conflict, adequate visibility at corners, minimal intersections, points of orientation, and change comes positive information rather than admonishment, guiding the user without creating an overbearing feeling of regimentation. These and related design moves are meant to minimize the need for signs and supervisory personnel (the traffic cop on every corner). Design does the bulk of the work in their stead.

8C. Safety

Design can also minimize the number of supervisors needed to ensure the safe pursuit of activities. And where such personnel are required, design can ease their chores.

The need for supervision can be lessened by reducing the potential dangers. Again, many physical hazards can be eliminated by attention to use-area relationships. For example, consider tot or casual areas where users, absorbed in benign pursuits, are oblivious to potential harm. Since their guard is down, they need extra protection from possible collisions caused by spillovers from adjacent areas. It follows that informal spots should be located away from or be well buffered from active areas. In addition, sport fields and the like should be oriented so that errant missiles, such as baseballs and Frisbees, will not fly into the midst of those whose guard is down. The most obvious measures for doing away with hazards should also be taken; such blind obstacles as drainage grates in the middle of ball fields should never be allowed.

We have said that the challenge of danger may at times be a part of the experience and should not be eliminated altogether. In such cases, design attention to both challenge and security should be bal-

Figure 4-54
A significant physical feature may serve as a point of reference where circulation directions are confusing.

Figure 4-55
Concrete "rumble strips" warn of an approaching intersection.

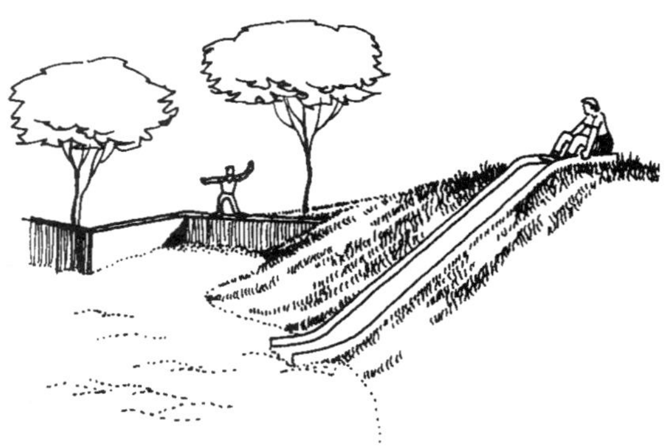

Figure 4-56
Slides, walls, and other play features can he designed for safety without minimizing their challenge potential.

Figure 4-57
A parental station should be located outside of the play zone yet remain within view of the children.

anced, as demonstrated in Figure 4-56. In the playground, walls to traverse might be raised high enough to provide a thrill yet remain low enough to avoid falls of bone-jarring proportions, and cushioning surfaces can be placed in the potential fall zone. Slides can be incorporated into natural slopes so that falls become tumbles rather than the straight-down variety you buy along with the standard iron-rung slide.

In high-risk areas where supervision is unquestionably called for, whether it be by parents or those on the payroll, the location of oversight stations and related sight angles should be carefully considered. Use relationships again play an important role. Similar facilities, for example a series of spaces used by tots, may need but a single station. If they are organized into a cluster, 360-degree vision is highly possible. If, however, such areas are scattered about the site, stations are needed for each space since there will be no one spot from which all play areas can be seen.

In addition to unobstructed visibility, physical proximity of oversight stations to play areas is desirable. Where physical proximity to all areas is not possible, certain priorities will suggest where the station should go. For instance, if a child who has slipped in the wading pool is plucked out quickly, no damage is done. But there is little one can do to avoid injury to the same child once it begins to fall from the top of a climbing device. If it can't be next to both, the station in this case, while within view of the climber, should be adjacent to the pool.

Since design is for people, the supervisors' comfort as well as their ability to perform should be taken into account. The station should be within view of but away from the actual lines of play to keep supervisors from being trampled. Details count a lot: shade, wind screening, and the like. And as illustrated in Figure 4-57, a low barrier between parents and sand pile is always appreciated, for it keeps the grit from their shoes and out of their reading material.

8D. Discouraging Undesirables

The problems of hard-core vandalism and perversion are extraordinarily complex matters for which there have yet to appear universal answers. In the search for root causes and solutions, parks sometimes serve as places for testing sociological and psychological theories and for the release of pent-up maladjustments and injustices. Significant along these lines are the works of a few social-action-oriented designers who have coordinated self-help projects in ghettos where residents and gangs have devised and constructed their own recreation areas on vacant land with locally available materials. One purpose is to spark pride through achievement, thereby combating deep-seated frustrations.

A related and very real concern of park administrators is the need to curb disruption caused by vandals and by sexual exhibitionists and potential molesters. Layout and design detailing can either assist or hinder in this.

For discussion purposes, vandalism has two sides, although in actual cases the lines are most likely to be blurred, so that control measures must be addressed to both categories. Relatively benign is the nuisance or "push over the outhouse for kicks" type of property destruction. Face it, folks. You did this, too. It is reasonable to assume that much of this damage is caused by people whose basic respect for property is momentarily pushed aside by a devilish impulse. A development that is neat and in good repair at all times inspires that respect. It appears worth cherishing and is therefore less likely to be violated than something that is beat up and broken apart to begin with. The latter invites further destruction.

By way of analogy, consider your new automobile. You take pains to keep from getting the first scratch on it. But once the first dent appears, the second does not seem to matter as much. And after the third or fourth, the car becomes in your mind just another beat up old "bomb" not worth worrying about. A park inviting this disdain becomes equally fair game.

But the park that does not have the first scratch on it is likely to be cared for, the respect it engenders in many cases squelching the momentary urge to cause damage. Accordingly, a park should be designed for easy maintenance so that with a minimum of effort, especially when that first scratch appears, it can quickly be put back into shape. The use of stains in lieu of paints, of appropriate surfacing where circulation will logically occur, and attention to such details as mowing strips are measures already mentioned in this regard.

The seemingly insoluble problems are in hard-core, habitually vandalized areas where the public's entrenched attitudes outweigh a basic respect for property. Where these cases are rampant, developments should first of all be designed for sweeping police inspection: potentially vandalized structures should be clustered rather than spaced about the site; views should be cleared from the street into the park; if the area is fenced, a minimum number of entry points should be located within view of well-traveled arteries.

To fence or not to fence is a subject of argument among law enforcement authorities. Some feel that fences with controlled gateways sufficiently discourage entrants bent upon destruction. Others maintain that fences impede capture; in the inevitable chase, the younger vandal vaulting fences with élan too often eludes the out-of-shape pursuer still struggling hand over hand up the barrier. They reason that the lack of a barrier equalizes the conditions of the chase, perhaps giving an edge to the police officer, who can drive into the park at any peripheral point and move at will over hill and dale without leaving the cruiser.

The structures themselves, whether they be buildings, signs, picnic tables, fireplaces, or drinking fountains, should be sturdy. And remember that the fewer the moving parts, the less chance of their being moved into the creek or someone's garage. A problem remains, however, which must be recognized by the designer. How do you provide sturdy structures that appear reasonably attractive, or at least humane? (See Figure 4-58.)

Plants chosen should be those that have a built-in chance to ward off the predations of vandals. Some say: Specify species with thorns; they can at least fight back. However, they also provoke lawsuits and can gouge the innocent. A more reasonable, although not always suc-

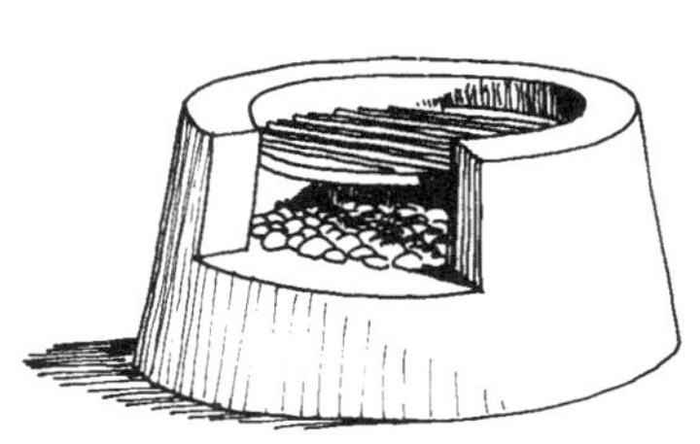

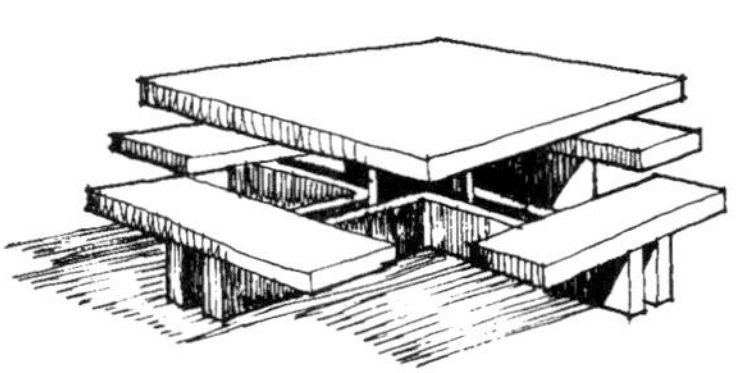

Figure 4-58
Sturdy yet attractive park furniture.

cessful, approach is to box young transplants with fencing until they are large enough to recover from a beating. It has also been noticed that individual specimens are always the first to go, whereas trees grown in clumps remain surprisingly free from destruction.

Many of the measures suggested to thwart vandals, such as opening the park to create sweeping views, are equally useful in discouraging deviants who would lurk to frighten or attack the unwary. Opening the park to view not only assists police inspection, it bares the act to the passing public as well, possibly making the psychopath think twice before harassing park users. Within the park, elimination of hidden corners near walls, building alcoves, and understory plantings are also useful, as is good night lighting.

The greatest attraction to marauders is the empty park and the lone person; the greatest discouragement is the park full of people. The designer can help fill the park with people; an experience-laden, efficiently functioning development should attract more people than a run-down, poorly organized facility. But the primary move comes from the park system planner and the recreation programmer. If a park is going to be alive with people, it should be located where a real need exists and activities of high public appeal should be slated. To further protect users, introspective pursuits can be allotted to areas which undesirables are less likely to frequent.

Understanding and Habits

The aim of this chapter as well as of the previous chapter on aesthetics is generating awareness. As with aesthetics, you are urged to establish an excellence scale for workability. Consider whatever aspects of size and quantity, orientation to natural forces, operation of machines, human comfort, administrative procedures, budget, materials, design details, freedom and control of use, circulation, safety, vandal-deviant proofing, and other aspects of park management come to your attention.

From this habit will spring knowledge that will help you rate your own parks and identify their needs, and give you some idea of how these needs might be satisfied. This is not to say that the clever schemes that work in one park will necessarily solve the problems of another. What knowledge gained through awareness will certainly do, however, is provide you with enough understanding of how design can satisfy demands per se so that you can judge the relative worth of any design solution proposed to meet your unique needs.

It should be obvious by now that there are many factors to consider in binding requirements into a workable design whole. Up to now, these have been presented in a somewhat scattershot fashion that may have left you lost amid a jungle of considerations, some of which appear to be at odds with others. Certainly, if our discussions are to help you develop an ability to analyze an intricate solution, a more organized framework for considering these issues must be developed.

In addition, while the principles and matters of concern have been held out as design substance or the critical issues that count toward the success or failure of most proposals, they are but categories and quite general at that. To be useful in evaluating a particular project, they must be made specific or be translated into issues that are relevant to the project at hand. To bring the critical issue of increased productivity further into focus, Figure 4-59 offers a catalog of suggestions.

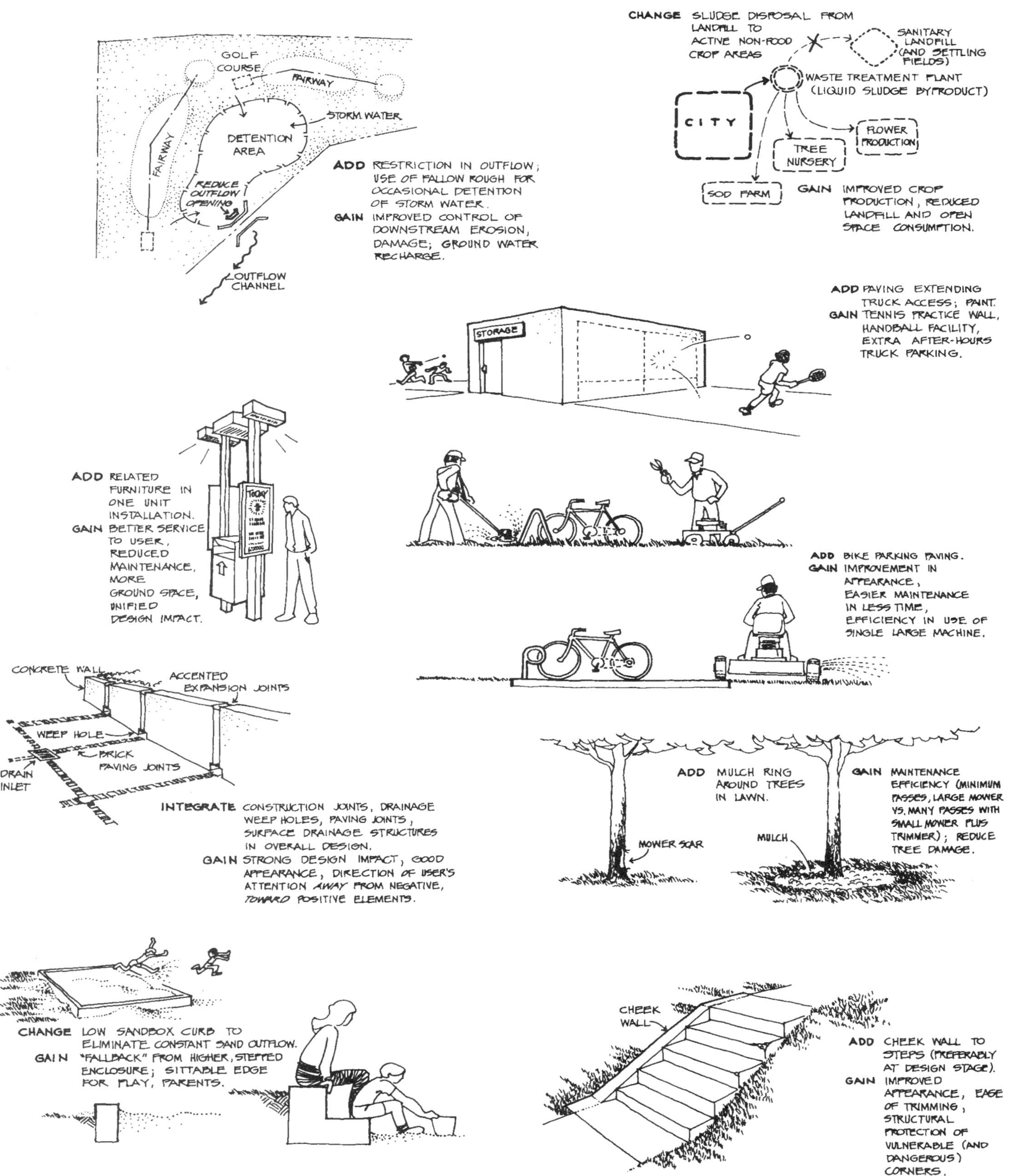

Figure 4-59
Major increases in productivity can result from minor design changes.

After a slight pause to discuss the mechanics of plan presentation, we will try to do this. First we will offer a process that designers use to organize their own minds, for they too need some systematic method for ferreting significance out of a jumble of possible determinants. The design process will then be used as a model which, with a few twists, will be converted into a critical procedure: a method that should turn generalities into specifics and allow you to strike at the heart of a design question while keeping the loose ends tidy as you proceed.

5 Plan Interpretation

Before we proceed to the planning process, we should discuss the plan itself and the symbols that have to be interpreted in order to grasp its meaning. Actually, there are several types of plans with which you should become familiar.

Plan Types

Before any scaled or refined drawing is done, a decision-making drawing is needed. Whatever ultimate size it represents, this will be a *schematic drawing* (Figure 5-1). These are sometimes called "bubble diagrams" based on the loose shapes (bubbles) drawn in a back-of-a-napkin style to represent elements or uses without any detail or true scale. They may represent relationships between elements. Or they may be schematics abstracted from the morass of detail about which decisions are made on a hierarchy of most important to least important without getting lost in the least important pile. They may serve as a courtesy to the reviewer. They are often used as backup material for verbal presentations or to accompany written reports that provide an explanation both of the plan and of the supporting data. Additional and subordinate *schematic plans* (might be used to illustrate circulation patterns, major relationships, or whatever else the designer believes need to be clarified.

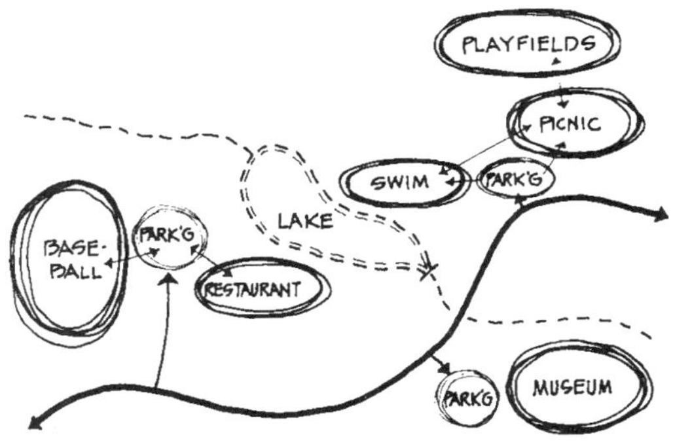

Figure 5-1
A schematic plan.

After planning decisions have been made and recorded with a schematic plan, the *master plan* (Figure 5-2 on p. 88) is created, which shows the essential organization of the park including commitments regarding circulation and major relationships: park to surroundings, use areas to site, use areas to use areas, and major structures to use areas. You will find that most documents specifically labeled master plan are for extensive developments devouring hundreds of acres or more and are drawn at a scale of 1 inch = 100 feet

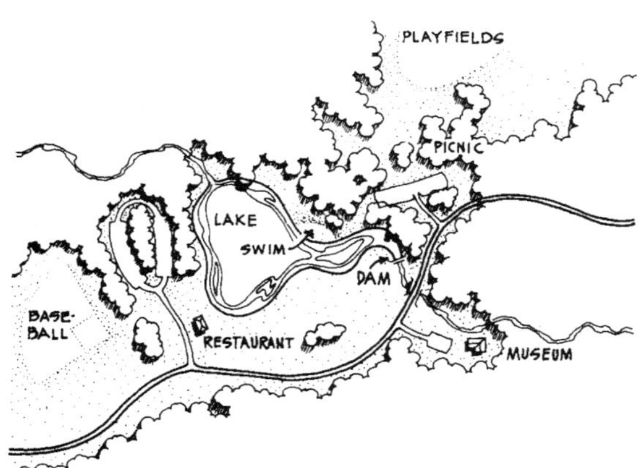

Figure 5-2
A portion of a master plan.

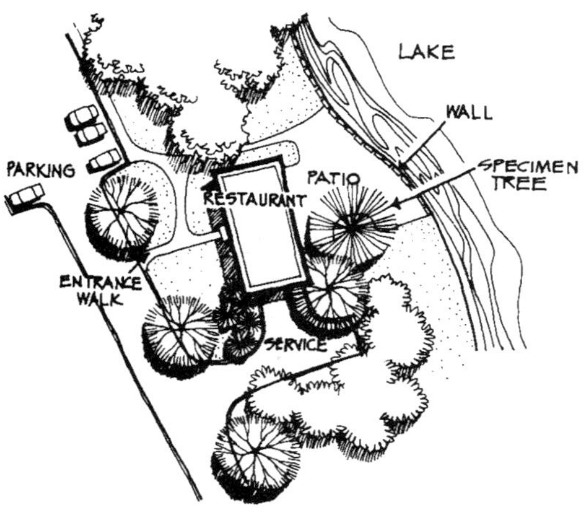

Figure 5-3
A portion of a site plan enlarged from Figure 5-2.

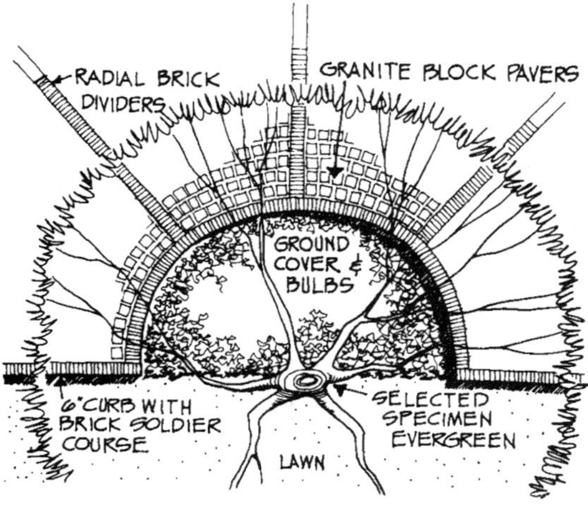

Figure 5-4
A portion of a detail plan blown up from Figure 5-3.

or 1 inch = 200 feet, the large scales enabling the entire layout to be seen on one sheet but limiting what can be shown to the most general. Even though design decisions are broad stroke at these scales, they still reflect much of the landscape architect's thinking about blending the new with the old, area sizes and quantities, orientation to natural forces, and use of existing site resources. What a large-scale master plan lacks is a full representation of proposed lines, forms, textures, and colors, all but the most major contour adjustments, the location of minor structures like light fixtures and drinking fountains, and the handling of other details too small to be drawn in.

These are left to be indicated on the *site plan* (Figure 5-3), drawn at a scale ranging from 1 inch = 20 feet to 1 inch = 40 feet. In addition to rendering more exactly the larger-scale commitments, the designer can also show aesthetic character, spatial framework, and the location of many minor structures on this drawing. The site plan can be a blowup of a portion of the master plan, or it can itself serve as the master plan when no drawing precedes it. The latter is usual if the area under study comprises about 50 acres or less, its moderate size being suited to a rather detailed solution presentation on one sheet of paper.

Where it is required that every fly speck including such details as curb widths, sign locations, and pavement textures be shown, a detail plan (Figure 5-4) drawn to a scale of 1 inch = 10 feet or less is in order. This can be an enlargement of a portion of the site plan or, if the area is quite small (a "tot lot" or bus waiting space, for example), can serve alone as a mini-master plan. Figure 5-5 shows an actual construction plan for the site.

There are other terms by which master, site, and detail plans are known. Development plan, master development plan, and, in some loquacious circles, master site development plan are a few which come immediately to mind. It is useless to suggest which label goes officially with which plan type inasmuch as there is no general agreement among designers, many using *master plan* to refer to what we have called a *site plan*, *development plan* for either site or *master plan*, and an infinite number of other combinations.

What is more important than haggling about titles is understanding what kinds of information each plan type provides, regardless of what the designer with whom you are working has chosen to call it. This will help you in two ways: If you know the degree of design commitment possible at each scale, you are most likely to concentrate your critical focus on relevant issues rather than waste energy searching for the location of benches on a 1 inch = 200 feet layout scheme; secondly, knowing what each plan can show will allow you to participate in decisions regarding what kinds of drawings are necessary.

What we have called a master plan is the first statement of development intent and suggests how all the pieces will fit. With a master plan available, each piece can be built separately with full confidence that, when development is finished, all pieces will knit and work well together. The master drawing is often accompanied by a *staging plan* that indicates how construction might be phased in instances where budget does not permit construction in one stroke. The master plan can also be considered a record document showing both the existing and proposed at any given moment. On it can be charted changes brought about by new demands that arise between plan creation and implementation. And because the plan provides an overview of the project, it is easy to spot the side effects of the changes, and to see what measures need to be taken to retain the original "fit" of the interrelated parts.

What we have termed the site plan gets us closer to construction, for the scale at which it is drawn allows the designer to pinpoint locations and draw rather precise forms. The site plan, often accompanied by detail plans of some of its parts, provides the basis for *construction plans* (Figure 5-5), which show measurements, material specifications, structural diagrams, and all the computations needed by contractors. It is also a guide for the *planting plan* (Figure 5-6), which is a type of construction drawing indicating to landscapers where plant material is to be installed.

All of these plans are interrelated, one laying the groundwork for the next. As you can see, there are two phases involved in seeing a job through to construction. The first is the stage of *ideas*, expressed in the schematic, master, site, and detail plans. These are the primary talking documents, discussed and modified a number of times before agreement is reached between client and designer. It is only after the ideas are agreed on that the designer can turn to the next, or *working drawing* phase, the construction and planting plans. Most nonprofessionals lack the technical background necessary for an in-depth review of the working drawings that will be turned over to contractors. Thus, it is during the earlier phase that your chief opportunity to participate as a critic occurs. Fortunately, this is a crucial period, for decisions reached at the idea level dictate most of what will be done in the working drawing phase.

Aids to Explaining Plans

A plan drawing is the usual form in which a site design solution is presented because it is the least time-consuming to prepare of the devices available to illustrate solution essentials. Unfortunately, it is a difficult document for the unschooled reader to comprehend, for while it is two-dimensional, its reality is three-dimensional. A scale model is a useful alternative. But unlike a plan, a model cannot be photocopied. It is also expensive to construct. Therefore, clients usually resign themselves to accepting a plan along with the mental gymnastics required to understand it.

To help a reviewer visualize the plan, the landscape architect often provides various types of sketches illustrating portions of the solution. These may be *elevations*, flat representations of objects seen on selected vertical planes (Figure 5-7); *sections*, slices through horizontal planes showing the goings-on above and below ground level (Figure 5-8); *eye-level perspectives*, three-dimensional sketches drawn as people might experience the development from chosen vantage points (Figure 5-9 on p. 90); or *bird's-eye perspectives*, overall three-dimensional views of the development as it might be seen from the upper story of an adjacent building (Figure 5-10 on p. 90).

While all these aid in putting the plan's message across, they can only show selected aspects of the solution. Even a comprehensive bird's-eye sketch has its shortcomings, for you cannot see behind and underneath many of its elements. Accordingly, you must examine the plan itself to decide most questions.

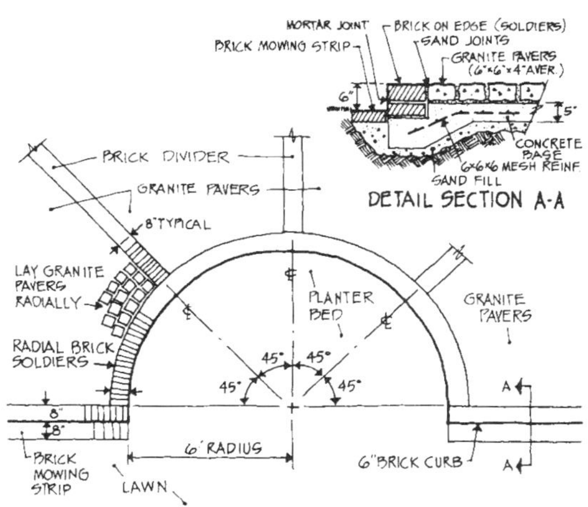

Figure 5-5
A construction plan for Figure 5-4.

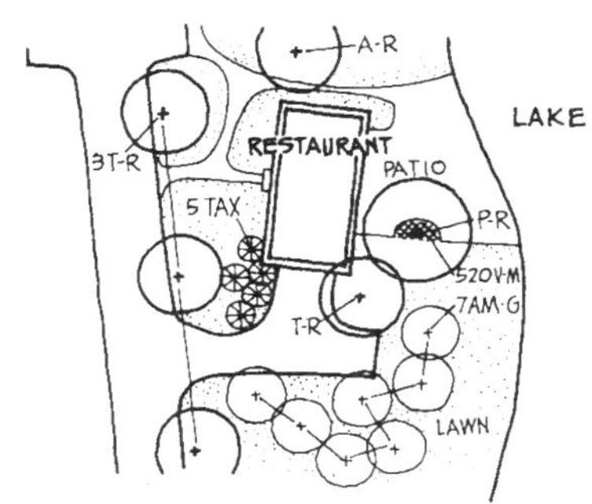

PLANT LIST

SYMBOL	NAME	SIZE	QUANTITY
A-R	Acer rubrum (RED MAPLE)	2½" CAL. B.B.	15
P-R	Pinus resinosa (RED PINE)	Specimen	1
TAX	Taxus cuspidata nana (NANA YEW)	24"-30" B.B	5
T-R	Tilia redmond (REDMOND LINDEN)	4" CAL. B.B.	4
V-M	Vinca minor (PERIWINKLE)	2¼" POTS	520

Figure 5-6
A planting plan for Figure 5-3.

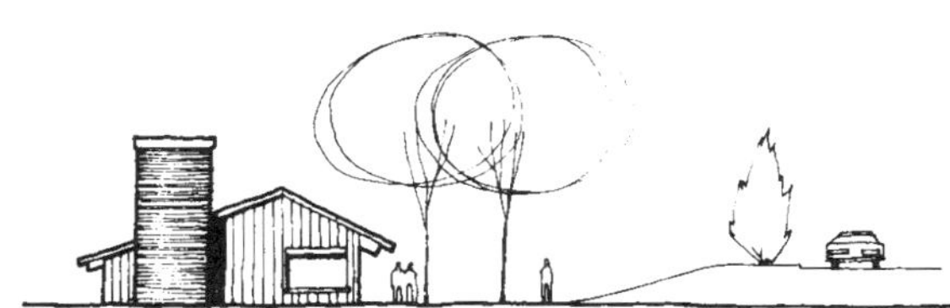

Figure 5-7
An elevation.

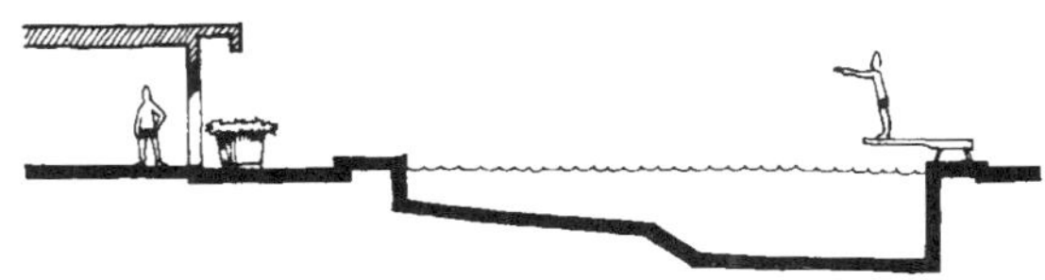

Figure 5-8
A section.

Figure 5-9
An eye-level perspective.

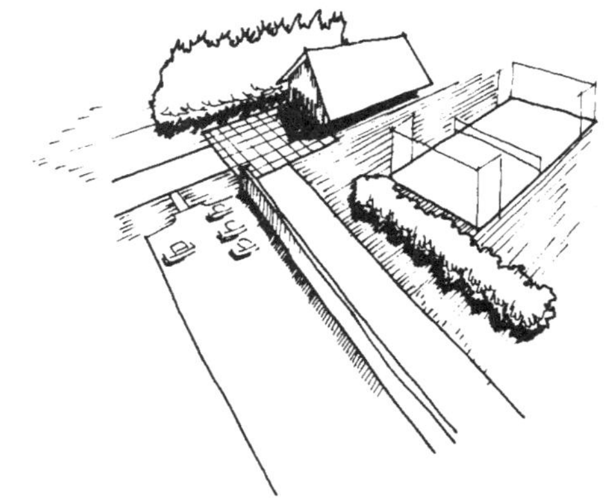

Figure 5-10
A bird's-eye perspective.

How a Plan Is Drawn

Using a variety of line weights and colors, the designer attempts to create several illusions that will ease the reviewer's interpretive chore (refer to plan drawings in chapters 6 and 7.) In this regard, a successful drawing is one that appears uncomplicated no matter how much information it includes. The designer accomplishes this by setting down visually important objects with heavy strokes and lesser elements with progressively finer marks. Hence, those objects that define space or form the skeleton of the plan (canopy trees, buildings, curb lines) pop out, while the other elements (pavement patterns, shrub bed lines, parking space lines) recede into the background. A simple guideline would be that the greater the vertical height of the element, the heavier the line. When there are several levels of visual strength each level can be read cleanly, whereas a monotone drawing would serve up the same amount of information as a sheet of unintelligible spaghetti.

The plan should also have a three-dimensional feeling about it to help readers make the transition from two to three dimensions in their minds. Edging plan features with shadow marks is a typical technique, but most of the job can be accomplished by the aforementioned line variations. Where this is well done, the plan's parts read in order of diminishing strength as if you were hovering over the site in a helicopter—which in a sense you are doing when going about your inspection of the drawing. The illusion thus created would also be analogous to the impact were you experiencing the actual development at eye-level. That is, buildings, trees, and other spatial definers would both read strongest on the plan and catch your eye first on the ground. Those elements just under the canopy or second to the buildings in height, such as fences and small flowering trees, which are drawn with the next heaviest line, would also follow the spatial definers in actual visual impact. Details such as grassy patches, cobblestone areas, and sidewalks would be drawn with the lightest lines, for in reality these would be the last elements to impose their presence upon you. Depending on the scale of the drawing, a curb line is a special case. For a larger scale drawing, 1″ = 40′ or 50′ the curb is a single, fairly bold line. For smaller scales, a curb may be represented by two parallel lines representing the face of the curb and the back of the curb. The face would be the bold line (the business edge of the pavement), but the back of the curb would be light, mainly shown to confirm actual space available.

The symbols on the plan that stand for items that the designer proposes to have installed can be labeled, but to facilitate interpretation they should also express something of the character of the items. Thus, designers might delineate trees as roundish bubbles with ragged edges so that they may be quickly distinguished from buildings drawn with unyielding lines and hard corners. They may symbolize brick with waffle-iron marks so you can immediately judge its extent in relation to the concrete shown on the same plan as a scored sweep and the grass depicted by texture stipples. They may also color the trees purple and crosshatch the water. Good luck to these dilettantes. They will have to spend most of their allotted time with you explaining what all those funny looking things on the plan stand for.

Contours

However, there are some odd-looking things on the plan that cannot be avoided if the form of the ground itself (both before and after construction) is to be described accurately. These are the *contour lines* wiggled over the paper to represent peaks, valleys, and slopes—the three-dimensional form of the land surface. (Note following letter references in Figure 5-11 on p. 92.) Contours are lines on the site drawings connecting points of equal elevation. In a way, these are the world's "bathtub rings," with the oceans imagined as the bath and the visible land as the tub. Likewise, if your park has a lake, it's the bath and the ground of the park is the tub. Imagine that in your lake is an island which is actually the tip of a huge cone extending down to the bottom of the lake. We'll drain the lake, lowering the water one foot at a time and marking the water line on the island's cone. Last, we'll take an aerial photo, straight down, of the empty lake. Where the cone stands, the photo will show the lines representing the one-foot levels of the receding water. They will look like the circles of a target, or bulls-eye, with the original island right in the center.

When applied to contours, the term *elevation* stands for a height above or below an assumed plane of reference or datum, the usual datum being sea level that is expressed as elevation 0. Thus, if a contour is labeled 90, the points it connects are 90 feet above the surface of the sea (A).

On most plans, dashed lines (B) represent the existing contours, while unbroken lines (C) indicate those proposed. This custom applies to hand-drawn site plans. In practice, using computer-aided design (CAD), it's more typical to find the existing lines drawn simply as lighter lines (fine dots, yielding a gray tone) and the proposed being darkest, using a solid stroke.

Visualization of the land configuration as expressed by contour drawings is a hard-to-come-by facility, but if the knack is acquired, the greatest obstacle to seeing the plan in three dimensions virtually disappears. Thus, a real effort to grasp the picture portrayed by these funny lines pays off handsomely. Here are a few technicalities you must consider in reading contour lines:

1. Every plan has a *contour interval*. The interval is designated in the plan legend (D) and stands for the vertical distance between adjacent lines. For readability's sake, every fifth contour is heavier (E). Therefore, if the contour interval is 5 feet, light lines connect, for example, elevations 80, 85, 90, and 95. The 75- and 100-foot contours are darker.
2. *Spot elevations* (× 92 on the diagram) are occasionally inserted to indicate critical points lying between the contour lines (F).
3. *The Contour Rule*: Water runs downhill, *perpendicular to the contours*.
4. The most significant land forms on a site are Ridges (G) and Valleys (H).

STOP!
WHY is G a ridge and H a valley?
They look exactly the same, like V forms.
We'll get back to this in a bit.

CONTOUR INTERVAL = 5' D
SCALE 1" = 100'

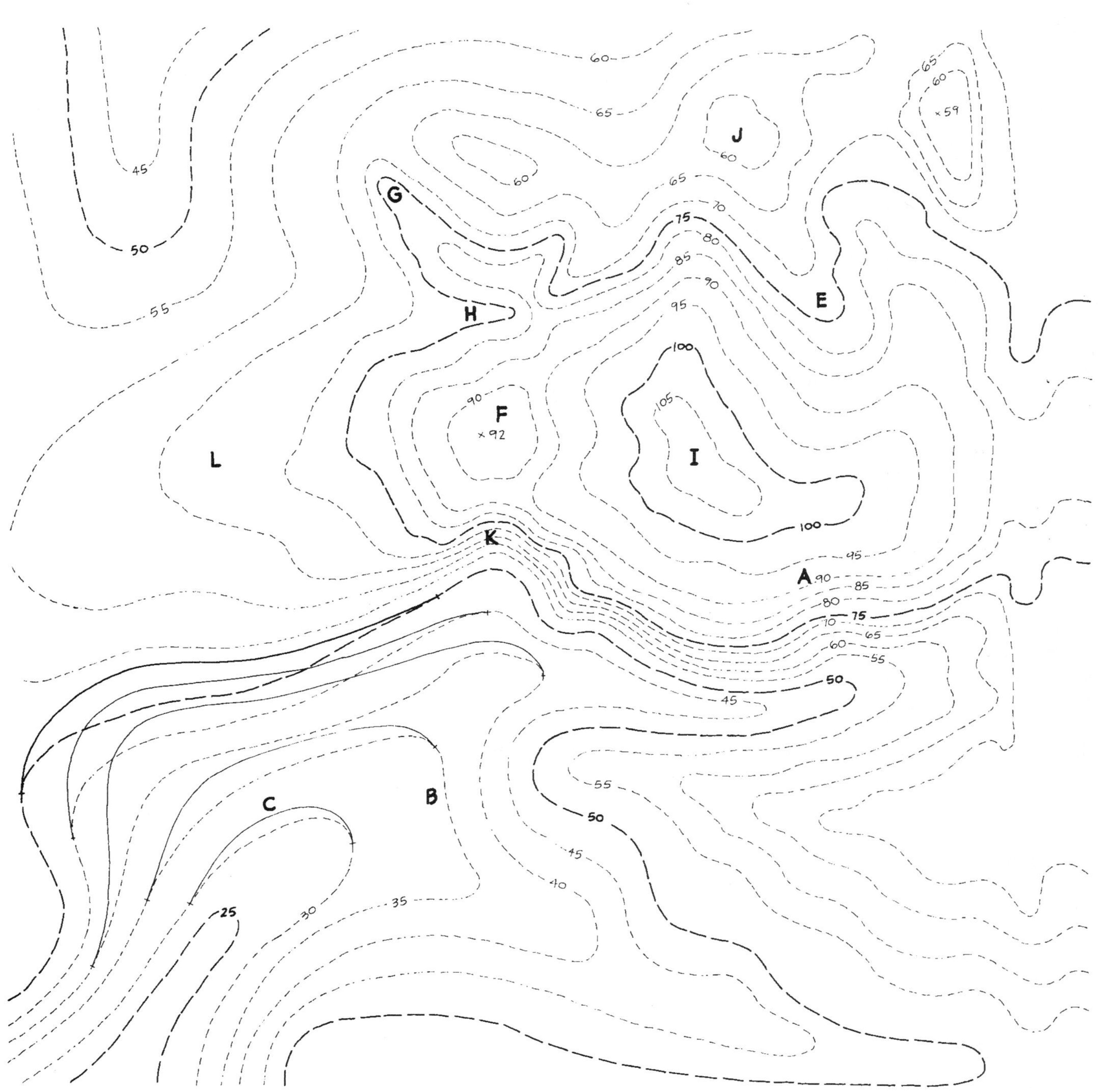

Figure 5-11
A topographic map.

5. A closed contour indicates either a summit (I) or a depression (J) in the ground. Usually a spot elevation within the closed contour will indicate the specific elevation to confirm it's up (summit) or down (depression) relative to the other adjacent contours.
6. The closer the contours, the steeper the slope (K); the farther apart the lines, the more gentle the pitch (L).
7. The steepest part of the slope is that which runs perpendicular to the direction of the contour lines (a corollary of the Contour Rule).
8. *Gradient* (slope pitch) is usually expressed either as a ratio of horizontal run to vertical rise (for example, 4:1), or a percentage. (A vertical rise of 25 feet per 100 feet of horizontal run would be a 25 percent gradient. More at Figure 5-14 on p. 94).

Now, about Those Ridges and Valleys:

Look at Figure 5-12.

This is a view of a model of the topo map shown in Figure 5-11. Each of the contour levels has been created as a single layer, 5' high, the same interval as the contour map lines. They are simply flat layers like terraces or rice paddies (in nature, the vertical edges would be covered with soil smoothed down over the forms).

You're looking at the topo plan as though it's tipped up with the bottom of the page closest to you and the top of the page in the distance. In this view, try to find the valley areas where the letters C and B are. Now look for the high point where the letter I is. Next find the ridge starting near B at elevation 50 and running to the right up to 70. Next find the swale near A, running down from 50, left and down to 25.

Now look at Figure 5-13.

This is the same model, but the view is shown with the right side of the original topo page tilted up facing toward you and the left side of the original page in the distance. In this view, try to find the same land forms. Also try to find G and H on the far side of the hill in the middle.

In both views, wander around and identify each of the elements listed above. Keep looking back and forth from topo map to topo model. This will help you quickly read the topography in the case studies which follow this chapter.

Apart from being rising, visually prominent landforms, *ridges* direct rainwater from their ridge lines (the highest points along the crest) down their sides to the *valleys*.

The sinking, lower landforms on the site, into which rainwater flows down the sides of the ridges, then turns to follow the valley floor or flow line downhill making streams, and rivers.

Contour Rule (reprise): Water runs downhill, *perpendicular to the contours.*

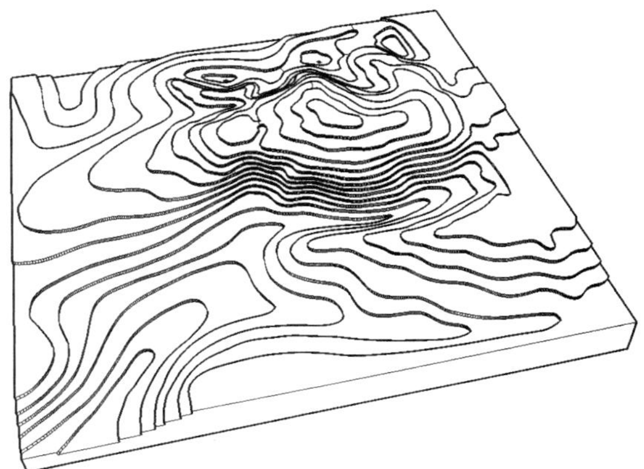

Figure 5-12
3-D view of topo seen from bottom of page looking toward top.

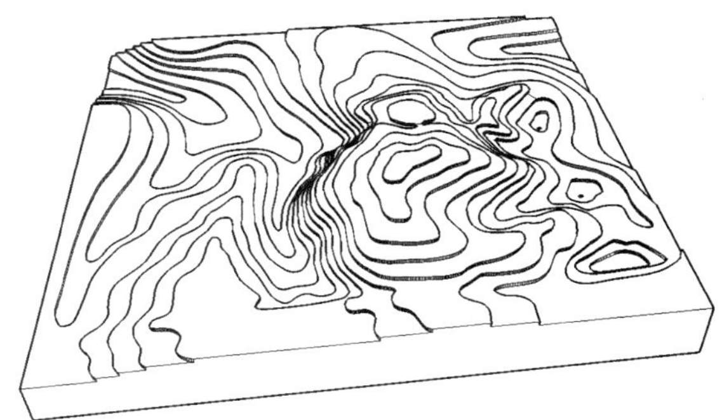

Figure 5-13
3-D view of topo seen from right side of page looking toward left side.

How Do You Read the Contours to Know Which V Is a Ridge and Which V Is a Valley?

Begin with the G, closest to the 75 contour. Look left and right in the V form. Right from the G, you'll find the 80 contour as another V. So that's up. Now left from the G, you'll see the 70 contour, also another V. That's down. Back at the G, look at the sides of the 75 V, Nearest adjacent contour is the 70, actually surrounding the 75, so G is higher than the adjacent contours. Imagine a line between the

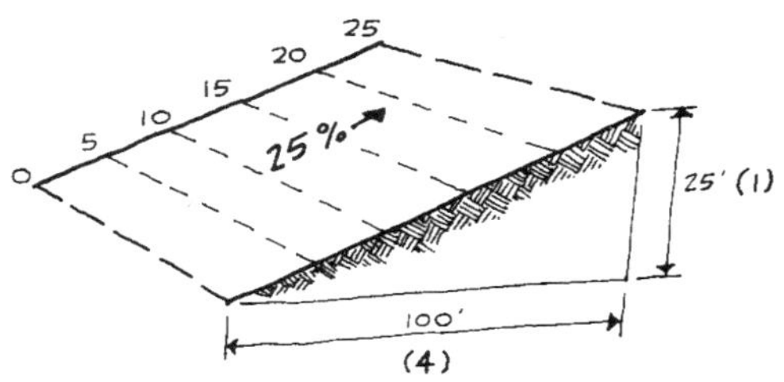

Figure 5-14
The slope of the land can be referred to in forms of a ratio or a percentage.

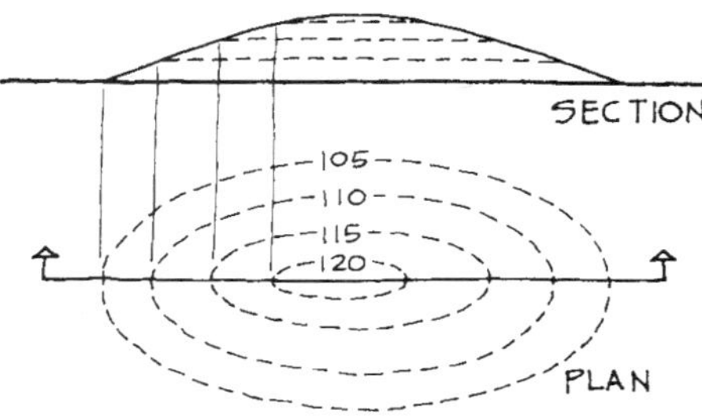

Figure 5-15
Visualized in cross section, these contours indicate a peak.

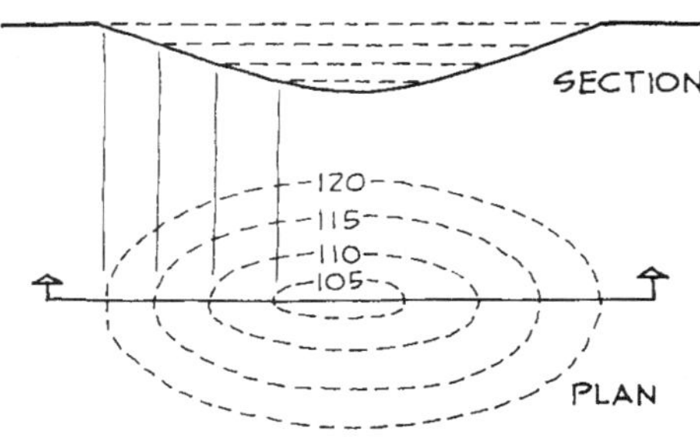

Figure 5-16
A cross-section shows that these contours depict a depression.

points of these 3 V forms, 70 to 75, to 80. Notice that on both sides of that center line the elevations are lower. If rain falls on the center line, it will run *away* from the center-line down to the surrounding contours. Must be a ridge! And the line you imagined from V to V is therefore a ridgeline.

Now go to the H. The 75 contour is closest to it also. Hmmm—another ridge? Look farther to the right of H, past the V of 75, and the next one is 80. Now back to H and look left; the next is 70. Same contours—what's the difference? Think another line, this time from the 70V to 75V and on to the 80V. This time, notice that the contours get *higher* on both sides of the center line, so the center line is the lowest point between the contours surrounding it. Must be a valley!

All too often in identifying ridges and valleys, the suggestion is something like "it runs in the direction the V points," referring to the ridge . . . or the valley. Then you have to remember which it is. Sorry. The only way to make sure is to do this simple process: Find the centerline of the Vs and see which end is high and which low. The water has no choice. Remember the Contour Rule: It's going to run downhill. Perpendicular to the contours. Always. Follow the numbers. Down the sides and down the centerline/flow line of the swale Vs. If down is toward the center line of the Vs, it's a valley (swale, flow line). If down is away from the center line of the Vs, it's a ridge.

The gradient can be easily computed from a plan where contours are shown. If you wish to express the grade as a ratio, first determine horizontal run by measuring the ground surface distance between several contour lines. This measurement can be taken with either an engineer's scale (marked off in decimals) or an architect's rule (scored in fractions) flipped to the side that represents the scale at which the plan has been drawn. Then multiply the number of contours found in the area of horizontal measurement by the contour interval. This gives you the vertical rise that is divided into the run measurement to arrive at the ratio. To express the grade as a percentage, simply count from the contour lines the number of feet the ground rises in 100 horizontal feet and place a percent symbol after the result.

Certain visualization exercises will help you get a feel for the relative steepness of the slopes represented by these figures and should otherwise give you a feel of the land. Develop the habit of doing cross sections or slices through the areas you wish to visualize. Do dozens on paper initially as demonstrated in Figures 5-15 and 5-16, and soon you will find yourself drawing them automatically in your head every time you spy a contour map or want to answer such questions as: Is that a peak or a valley? How rugged are those undulations? How flat is the surface proposed for the ball diamond?

Concurrently, you might run 100-foot measuring tapes out in the hall. Have one end held on the floor, raise the other the number of feet required for the grade percentage you wish to picture. For example, for a 3 percent slope, raise the end of the tape 3 feet. The "slope" of the tape will be the actual pitch of the ground whose contours compute to the figure you are using.

Begin to check out plans on the site. As you move through the development, picture yourself walking inside the plan. Eventually, you should be able to reverse this process so that every time you pick up a plan drawing, you will immediately climb in, walk around, and come away with not only a feeling for the landforms, but a three-dimensional impression of all the other design essentials as well.

6 Site Design Process

The ability to design recreation areas is similar to that required to solve any kind of land-use problem, the primary difference between a park design and, say, a subdivision design being the inputs analyzed compared to the outputs intended. Accordingly, you will find that most designers have worked their way into a park design specialty from a general education in landscape architecture, just as doctors receive a basic medical background before turning to brain surgery.

Development of design expertise does not occur overnight; it requires an inordinate amount of time and patience spent simply in doing: testing, discovering, and prodding latent talents to the surface. Much of the trial, error, and frustration that this entails is due to the fact that no one has yet invented a design cookbook containing procedures which guarantee results. However, many designers do use with some success a process that helps them sort out the facts, premises, and possibilities to arrive at a solution.

The process aids the landscape architect by systematically focusing attention on all factors that could affect the design's outcome and otherwise lends a semblance of organization to the attack. By illuminating relevant factors, it can also trigger flashes of inspiration. The process therefore bolsters both the rational and intuitive powers on which the designer must rely. While the process provides a framework, its ultimate value as a design tool is contingent upon the designer's ability to exercise those powers interchangeably within it.

A systematic approach to site design includes three phases. The first is a *survey*—assembling facts and data. The second is *analysis*—making value judgments about the effects of one fact on another. The third step may be called *synthesis*—weaving the results of analysis into a comprehensive solution to the problem. The steps may not be done strictly according to this chronology, for there is much feedback and interplay among them. In addition, the process must remain flexible in order to allow each designer to use his or her mental powers in the way he or she finds most comfortable.

Survey

Program Development

Each design phase has several parts. The first step in the survey phase is the preparation of a program expressing the early requirements of the project. Either handed to the landscape architect by the client, or its development delegated to the designer if the client does not have the means to shape it up, the program establishes goals to be accomplished. These might be set down as: *tangible items* (ten tennis courts), *capacities* (picnic facilities for 200 people), *physical benefits* (nonabrasive play surfaces), or *intangible gains* (as an educational institution, the school should have grounds that show respect for the natural environment). Agency policies whose implementation would depend upon how well the site was developed might also be included (a desire to collect entrance fees), as well as information regarding available *construction and maintenance funds*.

The program gives direction to the designer's thinking, yet it is always a flexible document, subject to modification as that thinking progresses. This is because both the answers and many of the questions remain unclear until the end.

In the book *Problem Seeking*, William Pena's small masterpiece on Caudill Rowlett Scott's programming techniques, Pena emphasizes the importance of controlling the "data clog" that results when the client provides too much information without any organization or priorities. The client needs the control offered by the designer's objective view of inflowing information: Is it important? Interesting? Irrelevant? Some data may seem very important to users because of current political or financial pressures yet have little long-term impact on the design. Other data may have to be obtained by, in effect, prying open apparently closed issues so that clients can recognize elements they take for granted and think of as unimportant but which have a direct bearing on the success of a project. For example, means of *increasing productivity* by intensifying or coordinating uses may not surface unless someone makes the effort to analyze the status quo of all facilities and then investigate both "*Why*?" and "*What else*?" In an urban park in an ethnic neighborhood, where the prevailing pattern is one of gregarious camaraderie along the street perimeters, conversation pits in the interior created by the original designer for adult users sit unused (and unmentioned by the users in the designer's information-gathering phase because, though they aren't a solution, they at least aren't a problem). Ask "*Why*?" and "*What else*?" Conversion to a safe play area could be a major program feature, but only if the opportunity is not overlooked.

Additional items might be suggested to take advantage of site potential discovered midway through the analysis phase: a skiing complex proposed to exploit the many steep slopes. Capacities may have to be adjusted due to site restrictions: fewer cabins planned to ensure that sewage distribution fields are not overloaded. In some cases, items may have to be eliminated entirely from consideration as unfeasible: swimming taken off the list because the water is found to be polluted.

As the designer begins to uncover solution possibilities, the program may also be augmented with criteria unique to the job at hand. Many will be spun from those principles and matters of concern found to be most relevant to the project. For instance, in the study of use-area relationships, the designer might realize that day users will be drawn to different attractions than overnight guests and conclude that facilities for the former should be separate from those for the latter. Additional directives might come from research into the technical requirements of the program items.

On counts other than the debated territorial imperative, in large measure the program should reflect the voiced desires of the people whom the park will eventually serve. However, on some sad occasions this is not the case, the list of facilities being drawn up in a vacuum by either the designer or administrator. As a result, the park may include only those things its developers like to do.

The citizenry becomes involved when many of the program items come from demand-study interpretations, the questionnaire route being quite satisfactory for most developments serving the diverse population that would be attracted to either a citywide complex or one located away from population centers. When development is proposed for a neighborhood, however, more direct public involvement in program preparation, including continuing face-to-face consultations with the designer, is highly desirable. One agency has created a full-time position for a public coordinator. The coordinator's responsibility is to know each neighborhood and the citizens living there, and to be their liaison with the park agency on all issues—design, maintenance, security, program, anything affecting their use of and regard for the parks. If not able to answer a particular question or resolve a specific concern, the coordinator brings the agency's designer or maintenance superintendent or other appropriate person directly to the place or person concerned to provide a hands-on response and, if possible, a proper solution.

This kind of involvement is especially significant for developments serving the urban poor, to whom the neighborhood is essentially the entire world and a major source of identity. To deny the residents direct participation, while outsiders ram through ideas they alone imagine to be cute, is both arrogant and patronizing. This is the way public alienation is created.

Inventory of On-Site Factors

After the program has been thrashed out, the designer begins gathering facts about the site, securing information from maps and personal inspections of the area. Such data could include the location of and other information about:

1. *Existing constructed elements*
   - *a.* Legal and physical boundaries, private holdings, and public easements
   - *b.* Buildings, bridges, and other structures, including those of historical and archeological significance
   - *c.* Roads, walks, and other transportation ways
   - *d.* Electric lines, gas mains, and other utilities
   - *e.* Land uses: agriculture, industry, recreation, and others
   - *f.* Applicable ordinances such as zoning regulations and health codes

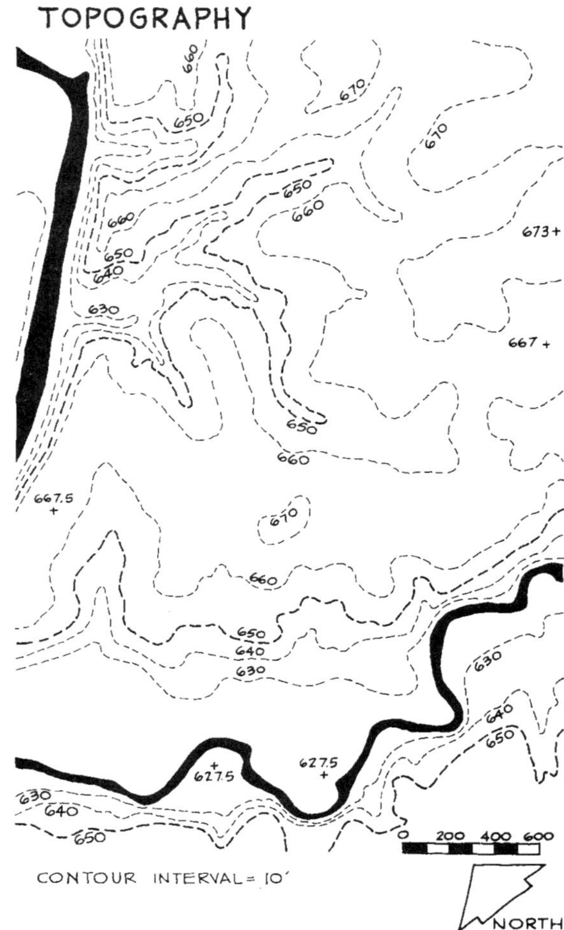

Figure 6-1
Site data—topography.

2. *Existing natural resources*
   a. Topography, including high and low points (Figure 6-1), gradients (Figure 6-2), and drainage patterns (Figure 6-3)
   b. Soil types, by name if available, for clues regarding ground surface permeability, stability, and fertility (Figure 6-4 on p. 100)
   c. Water bodies, including permanence, fluctuations, and other habits
   d. Subsurface matter: geology of the underlying rock, including existence of commercially or functionally valuable material such as sand and gravel, coal, or water
   e. Vegetation types (mixed hardwoods, pine forest, prairie grassland, others) and individual specimens of consequence (Figure 6-5 on p. 101)
   f. Wildlife, including existence of desirable habitats such as low cover for pheasants, caves for bears, or berries for birds
3. *Natural forces* (including both macroclimate as generally found over the entire site and microclimate characteristics or changes from the norm experienced in isolated patches)
   a. Temperature (air and water); especially day, night, and seasonal norms, extremes, and their durations
   b. Sun angles at various seasons and times of the day
   c. Sun pockets such as might be found in forest clearings; frost pockets that may be found in low places where the wind that sweeps away the morning dew is blocked
   d. Wind directions and intensities, both daily and seasonal
   e. Precipitation: rain, snow, and sleet seasons and accumulations; storm frequencies and intensities
4. *Perceptual characteristics*
   a. Views into and from the site; significant features
   b. Smells and sounds and their sources
   c. Spatial patterns
   d. Lines, forms, textures, colors, and size relationships that give the site its peculiar character
   e. General impressions regarding experience potential of the site and its parts

Inventory of Off-Site Factors

The designer must also accumulate information about the constructed, natural, and perceptual elements on the properties that surround or otherwise affect the site. These might include both existing and anticipated:

1. Land-use patterns
2. Stream and drainage sources
3. Visuals, smells, and sounds
4. Aesthetic character
5. Public utility locations and capacities
6. Transportation ways and systems

Recognize that the preceding information list includes data that *could* be collected but might not *necessarily* be collected. Learning

Figure 6-2
Site data—slope gradients.

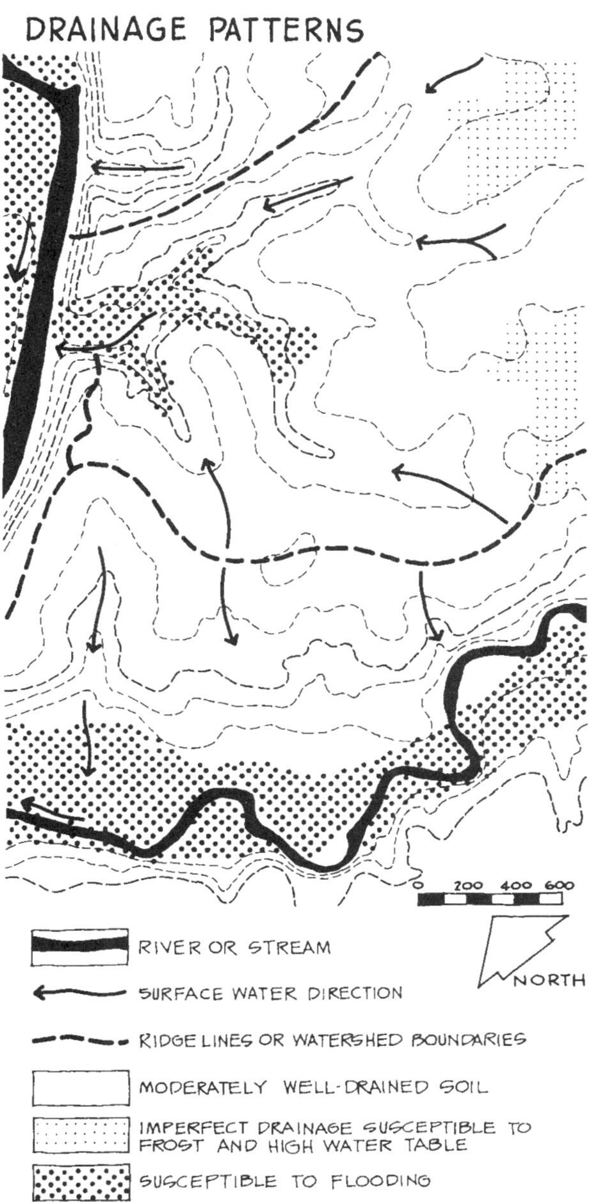

Figure 6-3
Site data—drainage patterns.

what is available has to precede a later determination of what is valuable. Each step in the survey phase begins in isolation, the first facts collected simply being those which are immediately handy. Soon all steps become intertwined, each giving direction to the others in order to turn general notions of what might be needed into specific requirements and to ensure that everything which could affect the project's outcome has been considered. The designer must shift back and forth between program and inventory. Program items suggest to the designer not only what information must be collected, but also what is inconsequential. The fact that a playground is being planned sends the landscape architect after data about sun angles and the peripheral traffic situation. It suggests little urgency to seek out a map showing the location of bear dens and pheasant cover. In a complementary fashion, as we have discussed, data garnered might make it clear that modifica-

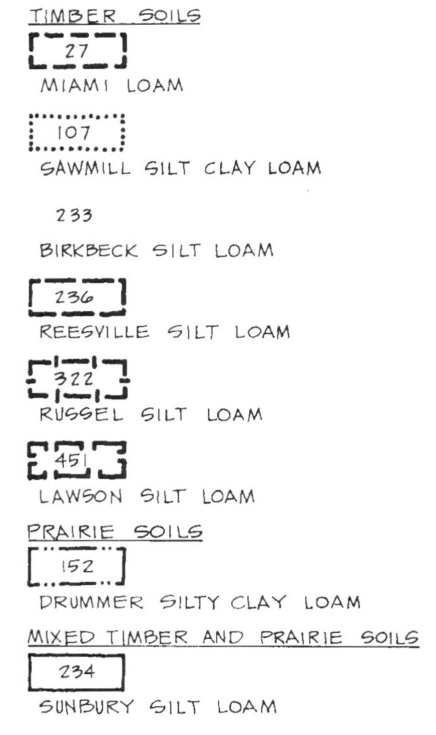

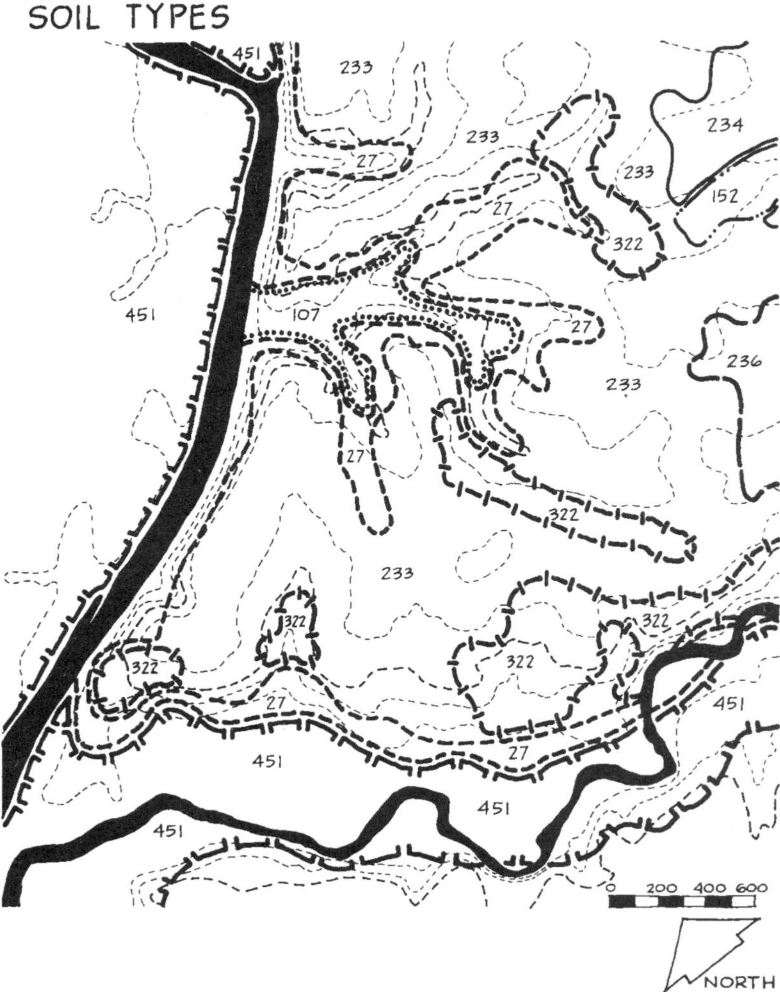

Figure 6-4
Site data—soil types.

tions are necessary in the program. The information gathered must be assigned some value in relation to program elements. The value may be a simple plus or minus, or data may be assigned to a numerical scale (for example, 5 or 10 = good or desirable, 1 or 0 = poor or undesirable). Unless the designer assigns specific values to the data, he or she may be misled by the data or find it hard to draw conclusions.

The recurrent question is: "Why?" What is the purpose of gathering this data? The target in site analysis is to use the information either to *find a place for a particular use* or to *find a use for a particular place*. Generally, with a program developed and uses identified, the focus of the analysis becomes the search for the right place.

To see how the survey phase and subsequent phases work, look at the accompanying case study, which has been simplified to eliminate confusing details. First, note in Figure 6-6 (on p. 102) the program comprised of items that we will assume have been agreed upon as meeting the needs of the park users. Relate the items to the drawing labeled *site analysis* (Figure 6-9 on p. 104) in order to see what categories of survey information the program has suggested. Then reflect on the diagrams and drawings in turn as each is explained in the following sections.

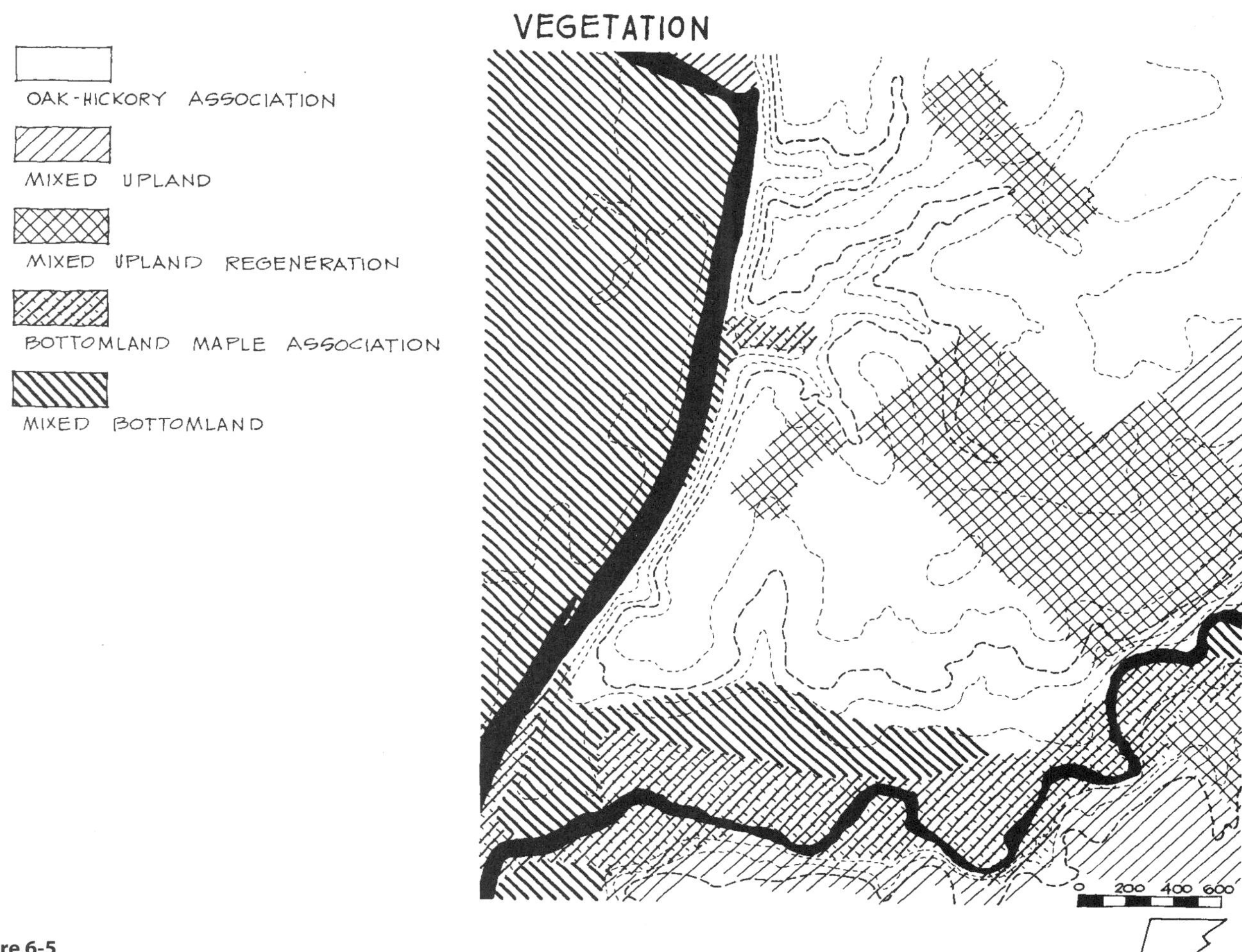

Figure 6-5
Site data—vegetation.

Analysis

Program Relationships

While data are being collected, design ideas are germinating, but they are held in the back of the designer's mind until a more comprehensive look at development possibilities has been taken. This begins by grouping program items in order to understand something about their interrelationships.

Relationship Diagrams

These interrelationships are then translated into diagrammatic form, the designer investigating how the major units might work well together and how circulation might be facilitated between them. Neither scale nor site information enters the landscape architect's mind at this point; the concentration is solely on functional relationships and lines of travel. Many combinations will be tried as the landscape architect attempts to work out the bugs of prior diagrams in subsequent ones. Finally, a scheme is developed which, in the designer's estima-

PROGRAM

ENTRY CONTROL BUILDING
NATURE CENTER
SERVICE YARD
BOAT REPAIR AREA
NATURE TRAILS
PICNICKING
PLAYFIELD
COMFORT STATION
SWIMMING
MARINA
ROWBOAT & EQUIPMENT RENTAL BUILDING
PARKING
CONCESSION
BATHHOUSE

Figure 6-6
Design process case-study program.

PROGRAM RELATIONSHIPS

1. NATURE CENTER
   A. TRAILS
   B. PARKING
2. PICNICKING
   A. PLAYFIELDS
   B. COMFORT STATION
   C. CONCESSION
   D. PARKING
3. SWIMMING
   A. BATHHOUSE
   B. CONCESSION
   C. PARKING
4. MARINA
   A. RENTAL BUILDING
   B. BOAT REPAIR
   C. CONCESSION
   D. PARKING
5. SERVICE YARD
   A. ENTRY CONTROL

Figure 6-7
Design process case study—program relationships.

tion, represents an ideal functional pattern for the program's major use areas. Although shown as separate drawings in Figure 6-8, the process could be more interactive, faster, and probably more productive if the bubbles—instead of being drawn—are paper cutouts rearranged, string bits laid between, and photographed. The photos could then be arranged on screen and rearranged with any simulation program.

Site Analysis (Listening to the Land)

Now the designer turns an analytical eye on the site, first compiling the inventory items on a topographic map. Those that lend themselves to graphic representation, such as vegetative cover, water, and soil, are expressed in sweeping patterns and graphically coded according to variety or condition (see Figure 6-9 on p. 104). This is done so that the character of the site can be read at a glance, allowing the designer to quickly spy areas suited for certain program items. For instance, landforms might be marked off in four colors or patterns: one designating slopes of less than 1 percent where drainage is always a problem; another standing for surfaces of 1 to 3 percent that may be suitable for all types of construction with little or no earthmoving; the third indicating areas of 4 to 9 percent grade, the land suited to building and roadway installation with moderate grading; and the fourth showing slopes of 10 percent or greater that will require major site adjustments if built upon. Breakdown increments will vary with each project, depending on the slope requirements of the program units.

To help visualize the landforms represented by the topographical base map, Figure 6-10 (on p. 105) raises the contours in layers proportionate to the contour intervals (as was done in Chapter 5). The relatively flat, easily developed land areas are more obvious. And the steeper areas show the relative difficulty of trying to site active-use areas on these slopes.

In addition, natural and perceptual influences like wind directions and noise sources are illustrated with bold symbols, and other considerations are set down in note form on the map. The latter might be directives the designer has established internally according to the ideas that have been germinating in the back of the mind: "Save rock outcropping." Or, because by now there is a feel for the program requirements: "Good place for entrance." Managing and presenting this kind of information is a challenge discussed later in reference to *computer support*.

In the analysis phase of the design process, the designer strives to gain a full understanding of program requirements and an intimate knowledge of the limitations and potentials of the site before beginning to make concrete decisions. This phase also serves to flash possibilities for solution in the designer's mind, alternatives founded in the logic of the analyses being made.

Sustainable design or common sense, whatever the term chosen, is the guiding principle in play as each element or condition is considered in this process of analysis. The designer must constantly *listen to the land*. Seeing a grove of very old trees growing on a sun-washed, fairly steep gravelly slope, the designer should hear the site saying, "More of these will do fine here." The designer who then chooses, for the same area, to add a species with color and texture that seems interesting and beautiful (even though that species likes low, flat terrain with a moist, rich soil) is site-deaf, resulting in (a) a nonsustainable result that has to be constantly and expensively nursed along to survive; or (b) a total loss! The same consideration has to be made of

RELATIONSHIP DIAGRAMS

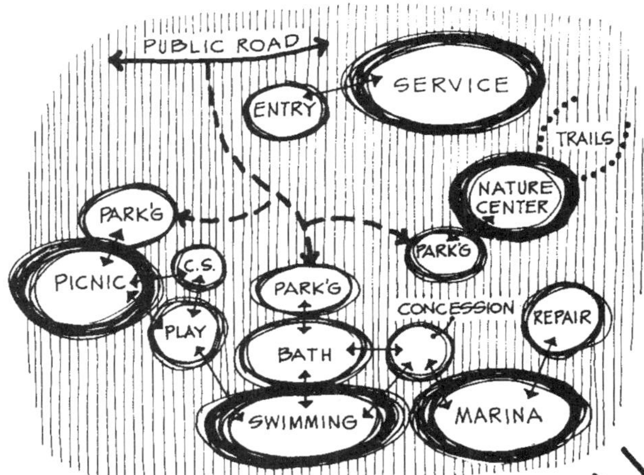

- \+ ENTRY CONTROL NEAR PUBLIC ROAD.
- \+ PARKING - SPLIT, DOESN'T PENETRATE ACTIVITIES.
- \+ ONE CONCESSION FOR SWIMMING & MARINA WHICH SHOULD BOTH BE NEAR WATER.
- \+ PICNIC & SWIMMING SHARE PLAYFIELD.
- − MARINA NOISE & DEBRIS CONFLICT WITH SWIMMING.
- − PICNIC DIVORCED FROM NATURE CENTER.
- − NATURE CENTER WEDGED BETWEEN NOISY SER ICE AND REPAIR.

- \+ BATH-CONCESSION COMBINED TO SEPARATE SWIMMING & MARINA.
- \+ PICNIC & NATURE CENTER PARKING COMBINED.
- \+ SWIMMING & MARINA PARKING COMBINED.
- \+ REPAIR & SERVICE (NOISE & MESS) COMBINED.
- − PICNICKERS GO THRU N.C. AND TRAILS TO SWIM & BOAT.
- − ROADS CUT PICNIC FROM BOATING.

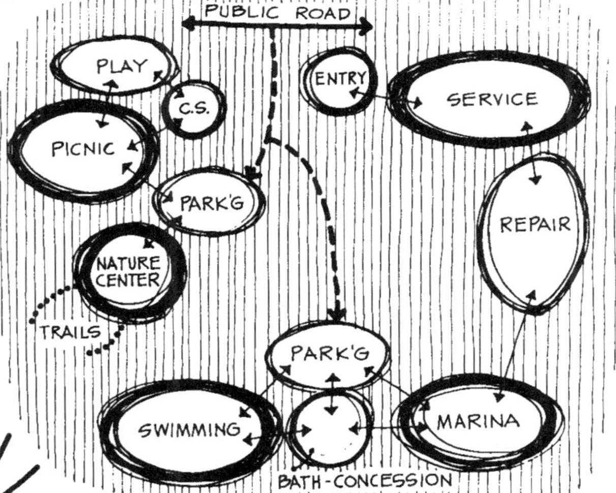

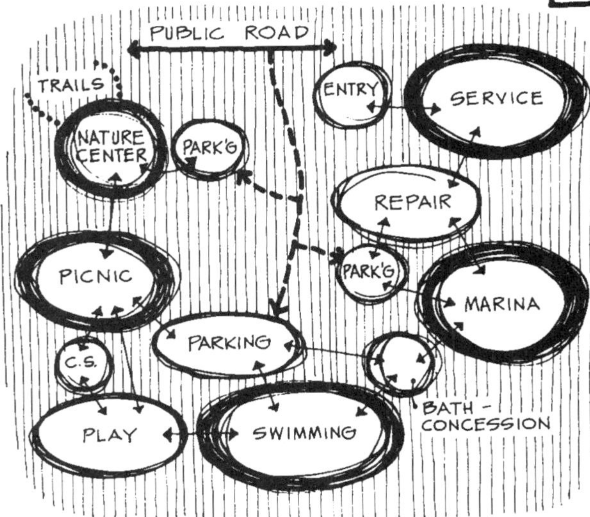

- \+ PARKING - SPLIT, DOESN'T PENETRATE ACTIVITIES (PEOPLE DON'T CROSS ROADS TO GET TO ANYTHING).
- \+ N.C. ISOLATED BUT CONVENIENT TO PICNIC AREA WHICH WILL PROVIDE MAJORITY OF VISITORS.
- \+ BATH-CONCESSION SEPARATE MARINA & SWIMMING.
- \+ PLAY SEPARATE BUT USABLE BY BOTH SWIMMING AND PICNIC.
- \+ MARINA, REPAIR & SERVICE TOGETHER - NOISE ISOLATED FROM REST OF PARK.
- \+ ENTRY CONTROL NEAR PUBLIC ROAD, FIRST CONTACT.

THIS IS IT!

Figure 6-8
Design process case study—relationship diagrams.

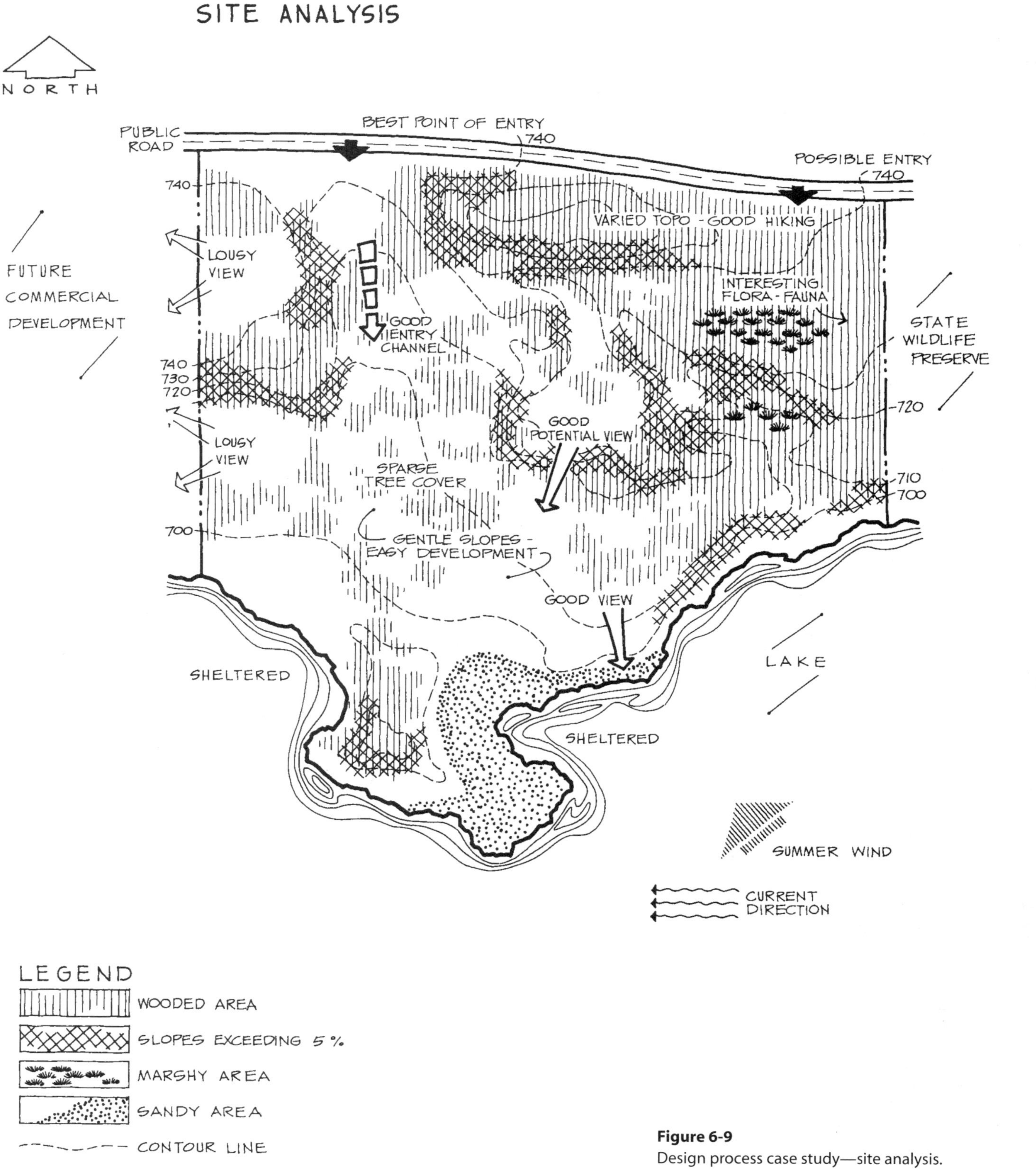

SITE ANALYSIS

Figure 6-9
Design process case study—site analysis.

construction materials chosen and landform manipulation by grading. *What does the site say?*

As in the survey phase, analytical steps are first taken singly but soon swing into a back-and-forth effort, with a growing knowledge of program needs directing the search for particular site qualities, and a rising feeling for the site illuminating what can be done to satisfy the program's demands. Concurrently, both research and criteria reevaluations are proceeding so that, when the time comes to put the pieces together, the designer knows full well what tests a successful synthesis must pass.

The process described thus far is proper and reasonable. A proper conclusion is assumed to be taking further steps in the design for the site. However, be mindful that one conclusion could be that *this is the wrong site*! If there are too many conflicts with reasonable development, the best step could be to back up and look for a better fit. This is unquestionably a better time than when tons of earth have been moved and yards of concrete poured. *Listen to the land.*

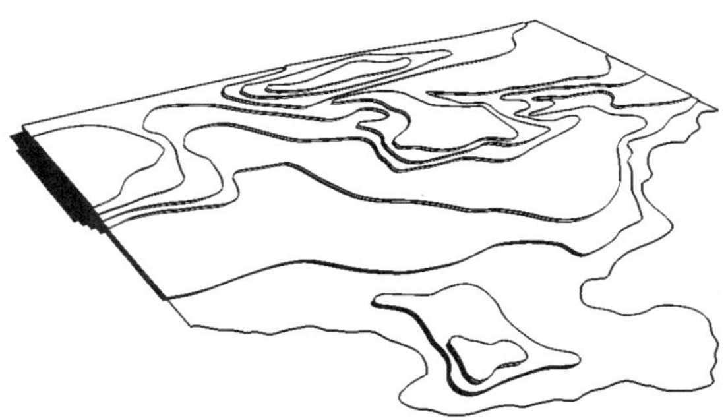

Figure 6-10
Model illustrating 3-D landforms from topographic base map of site analysis

Synthesis

Design Concept

Up to now, one issue at a time has been placed in front of the landscape architect in a progression of interrelated complexity. Proceeding in such a fashion, the designer is less likely to be panic-stricken by the magnitude of the problem than if attempting to address the entirety of the project at the onset.

Once the designer has a good handle on required relationships and site influences, the synthesis phase is opened with an attempt to fit the ideal functional diagram to the site. What the designer reads from the diagram and tries to visualize on the ground is a functional essence rather than a literal image; that is, a diagrammatic sketch which shows one use unit next to another does not demand that one must appear to the right of the other on the final plan. Rather, it offers the more general suggestion that they must be adjacent on the site. The designer is therefore free to reverse, shift, rotate, warp, or otherwise manipulate the diagram elements so that all use units end up on desirable portions of the site in a pattern that retains the essential relationships of the abstract diagram.

What takes place in the designer's mind during synthesis is impossible to describe except to say that it entails a series of impulsive reactions to the conditions of the problem as they unfold. In no particular order, save one with which he or she feels at ease, the designer may start by selecting a perfect location for one use area, work circulation to it, then roll the other units from it. The designer may relocate areas whose initial positioning violates the sense of the relationship diagram or revise the whole to ameliorate the negative effects of the relocation, and then turn to refinements that satisfy a single criterion. Gathering steam, the designer begins reacting to all factors at once, working simultaneously among all the solution aspects until at last the whole is visualized (see Figure 6-11 on p. 106).

The product is a skeleton concept, much like an outline for a book, only presented in graphic form. Use units are illustrated in approximate size and form where they belong on the site. Traffic channels are indicated. The spatial structure is set. Additional work

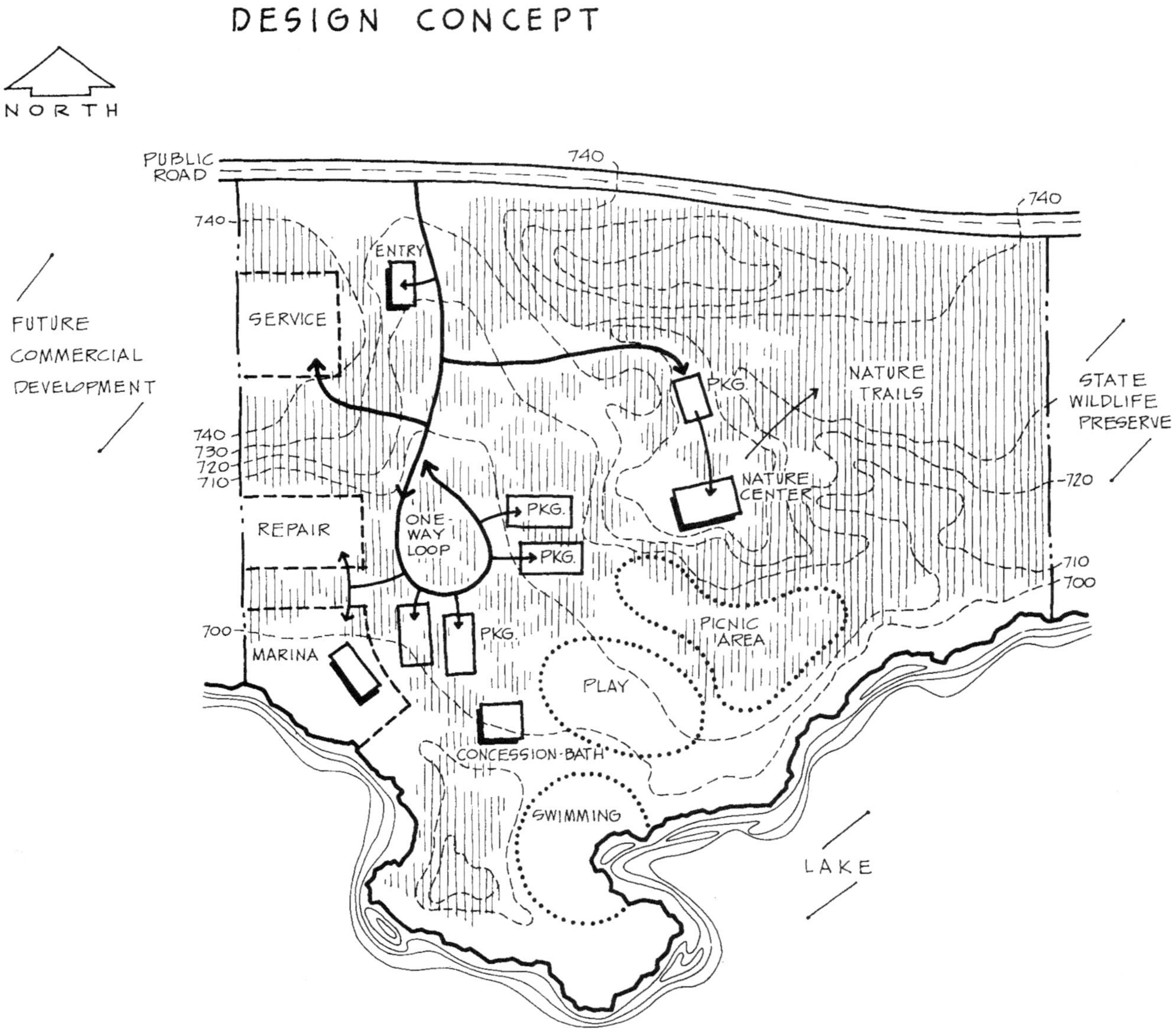

DESIGN CONCEPT

PUBLIC ROAD
740
740
740
FUTURE COMMERCIAL DEVELOPMENT
SERVICE
ENTRY
PKG.
NATURE TRAILS
STATE WILDLIFE PRESERVE
740
730
720
710
NATURE CENTER
720
PKG.
ONE-WAY LOOP
REPAIR
PKG.
710
700
PICNIC AREA
700
PKG.
MARINA
PLAY
CONCESSION-BATH
SWIMMING
LAKE

RELATION OF USE AREAS TO SITE

1. NATURE CENTER
- GOOD VIEW
- APPROXIMATE TO DIVERSE NATURAL INTEREST AREA
- ISOLATED BY TOPO FROM OFFENSIVE TRAFFIC & NOISE

2. PICNICKING
- FLAT SITE
- STABLE SOIL
- CANOPY TREES
- VIEW OF WATER, YET AS SAFETY MEASURE, SEPARATED FROM LAKE BY TOPO
- IN PATH OF SUMMER BREEZES FROM LAKE
- RELATED PLAY AREA IN FLAT, TREELESS SPACE

3. SWIMMING
- SANDY SOIL
- SHELTERED COVE
- MAXIMUM SOLAR EXPOSURE

4. MARINA
- SHELTERED
- APPROXIMATE TO COMPATIBLE COMMERCIAL DEVELOPMENT
- FLAT SITE
- STABLE SOIL
- RELATED REPAIR AREA FLAT AND TREELESS, UNAFFECTED BY POOR VIEW

5. SERVICE YARD
- FLAT, CLEARED SITE, UNAFFECTED BY POOR VIEW
- STABLE SOIL
- SEPARATED FROM PUBLIC AREAS BY TOPO

6. ROADS & PARKING LOTS
- FLAT TOPO AVOIDING EXCESSIVE CUT & FILL
- STABLE SOIL
- SPARSE TREES PROVIDE COVER YET ALIGNMENT AVOIDS REMOVAL OF VEGETATION
- EXISTING SPATIAL CHANNELS

Figure 6-11
Design process case study—design concept.

remains, for the skeleton must be fleshed out with myriad details, but these will supplement rather than materially change the major commitments made at this point.

STOP!
Revisit the question, "Is this the wrong site?" Could be. Nothing is working out happily. Maybe the other sites are better? Which other sites? (Hopefully not meaning "We don't HAVE any other sites.")

If a program has been developed with solid public support, finding the right site is the necessary first step. Either department staff or consultants should be charged with developing criteria for site selection and gathering survey data on a few sites to be evaluated. The steps outlined, from inventory through analysis, should be performed on the potential sites. The suitability of each site should be ranked and the selection made.

Alternatively, the agency may already own a particular site, or one is being given. In this event, the program elements could be listed as desirable potentials and the site could be evaluated for suitability to the possible uses. A quick look forward at Figure 6-13 (on p. 109) shows each use scored for suitability, from which the likely list is developed. With the list of possible uses in hand, the relationship diagram alternatives would be developed more by discovery on the site map than in the abstract as diagrams.

Refined Plan

Within the concept framework, the designer proceeds to make precision adjustments, add minor use areas and structures, and do whatever else is necessary to give the solution its finished form (see Figure 6-12 on p. 108).

Final Plan

The work is now drawn up as a *preliminary plan* ready to be reviewed by the client. A single refined concept may be presented if the landscape architect feels that the problem can be solved in only one manner. A number of alternatives may be trotted out if the designer has either evolved several equally desirable relationship diagrams or discovered that a single diagram can be fitted to the site in a variety of ways.

During the review session, the designer explains the reasons behind the moves while the client darts a question here and there, mulling over the ideas generated by the drawings. A successful review ends with agreement about the proposal, usually contingent upon the incorporation of modifications requested by the client. Back at the drawing board, the revisions are made and the plan is finalized.

Computer Support

As we have seen, the site analysis generates criteria that guide the siting of the program uses. These criteria are established by relating the program items to the inventory information collected on the site. Organization of the inventoried information can at first seem like an overwhelming task, both in terms of raw volume of data and number of possible interrelationships. In the case just discussed, for example, fourteen different uses are considered in relation to more than a

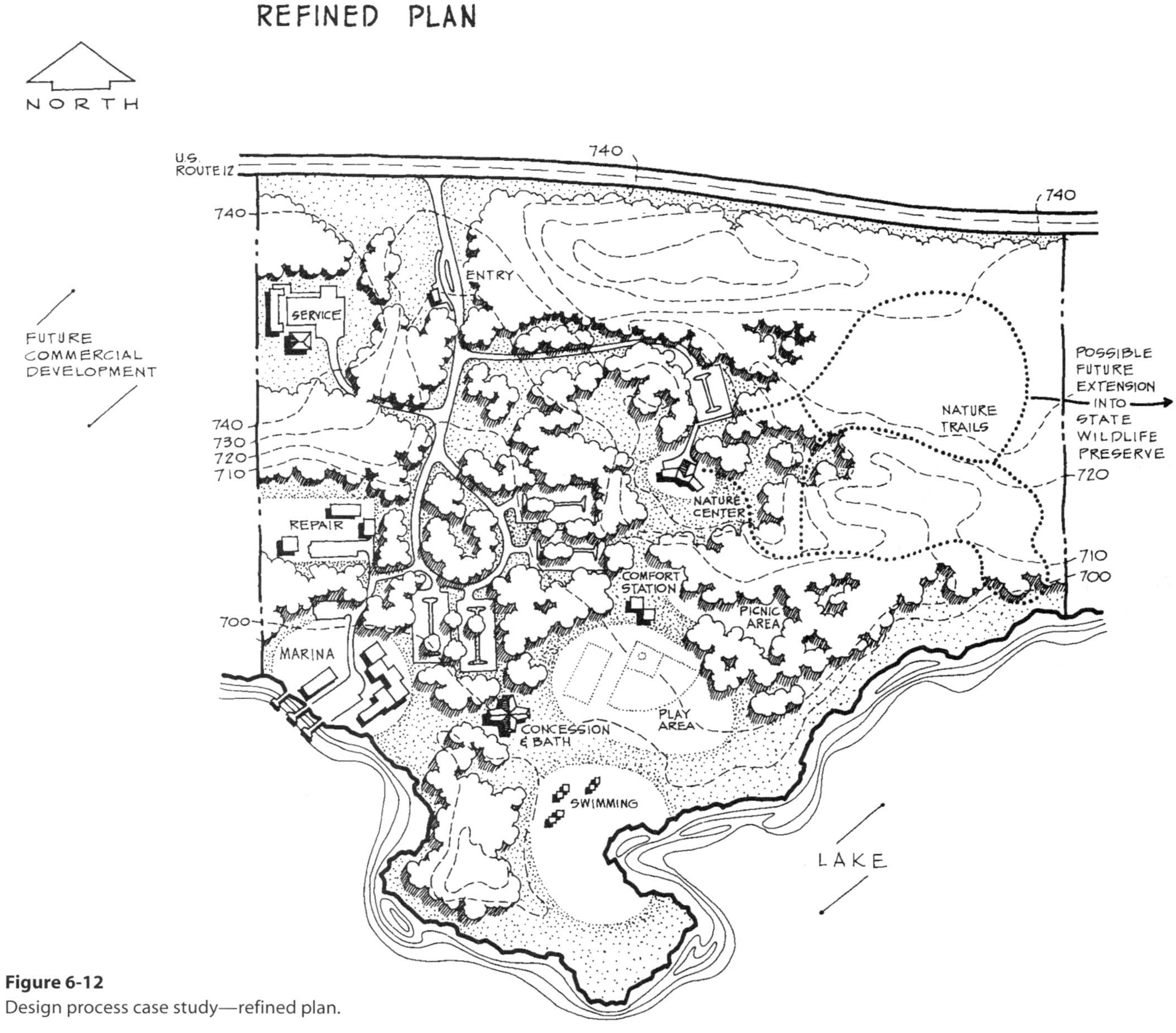

Figure 6-12
Design process case study—refined plan.

dozen different criteria, including steepness of slope, vegetation cover density, and highway access.

The application of data-manipulation technology offers the only sensible means of dealing with Pena's data clog difficulty. Using software created for management and display of data (usually a spreadsheet or database management program), a matrix can be generated showing interrelationships of all the uses being considered and all the criteria assembled so that uses with similar requirements can be identified and grouped easily (Figure 6-13). This consolidation can make the selection of suitable sites easier. The criterion most limiting to all uses is apparently steep slopes. Plotting these data on the base map would immediately eliminate consideration of many areas for an entire block of uses.

Consequently, data such as topographic lines may be shown on the "topography" data map as part of the *inventory* phase. Based on that inventory, however, the more useful information becomes part of the *analysis* phase where patterns on the site's topography are identified as "slope gradients" with the slopes shown (such as 2–4%). Further support for the analysis would have the additional qualifier "too steep for field games" added to the percentage descriptors.

Dense vegetation is another major criterion that can be plotted, further narrowing the array of site choices. Like the topo information, vegetation data would be described on the *inventory* map as "mixed upland" but later on the *analysis* sheets as "too dense for small-court games." Soils information appears in the *inventory* under the usually exotic names of the local soil types but in the *analysis* phase with its true influence noted as "structurally stable" or "will not support athletic turf." The *inventory* information about drainage patterns only becomes useful when it appears in the *analysis* as "area of severe erosion" or "seasonal flooding," serving to guide land-use decisions affecting the park's functional limits. These more descriptive criteria can then be included in a matrix column for immediate graphic evaluation of each potential use against possible site location.

Reversing the telescope, the data management matrix can be used to display isolated criteria for single uses. Instead of analyzing a single site for many uses, this kind of application compares several sites for a single use. In such a situation, the question addressed could be: "Of these four sites for potential acquisition, which would be best for field-games development?" The output would include the focused criteria list as shown down the left of the matrix in Figure 6-14. However, the facility-type column heads would be replaced by the roster of sites showing their respective *scores* for each of the criteria. The selection process would then at least start with a ranking of known data, to be confirmed by more detailed physical observation of the properties. In this situation the site having the highest score (site D) would be most suitable.

If the computer's only function in the site analysis process was to provide these matrix displays, it would have to be considered merely as an expensive assistant. The dramatic importance of this technology is in the step beyond the first presentation of collected data. The game of "What if?" can next be played simply by changing any of the criteria. What if each slope could be increased by 2 percent? For the individual park site, all affected uses immediately are displayed with revised relationships. Some uses might be possible in areas previously designated as off limits. For the specific use, the several sites being evaluated might suddenly have a new rank order with a consequent change in purchase priority. Interesting. What about a 3 percent increase? What about *decreasing* by 2 percent? The instant display of results is a vitally important adjunct to the effort to make the best decision about resource utilization; it guarantees that the maximum possible number of alternatives has been considered and the minimum number of surprises will be encountered.

Depending on the complexity and size of the project, the plan-related presentation of the inventory and analysis parts of the planning process may require a graphic treatment as a series of overlays to a base map, each overlay representing just one element of information. In the case of the rock outcropping noted earlier, this item might in fact show up on a particular layer of information described as *Significant Elements* with evaluations noted as positive or negative by degrees. The evaluation of a potential entrance noted earlier might be a part of a particular layer of information related to *Traffic Issues*, if access and circulation are significant influences on the master plan-

SPREADSHEET FOR SITE ANALYSIS—Project 6/85
Suitability Criteria

FACILITY TYPE		I	IIA	IIB	III	IV
SLOPES	0–2%	x	x	x	x	x
	2–4%	x	x	x	x	x
	4–10%	x		x	x	x
	10–20%	x				
	20%+	x				
DRAINAGE	Good	x	x	x	x	x
	Fair	x		x	x	
	Poor	x				
SOIL	Good	x	x	x	x	x
	Fair	x	x	x	x	
	Poor	x				
VEGETATION	Good	x		x		
	Fair	x		x		x
	Poor	x	x		x	x

FACILITY TYPES INCLUDE:
I. Nonconstruction: nature trails
IIA. Minimal construction: playfields
IIB. Minimal construction: picnicking
III. Moderate construction: roads, parking, service yard
IV. Intensive construction: all structures

Figure 6-13
Typical spreadsheet format for preliminary computer-based site analysis.

SPREADSHEET FOR SITE SELECTION—
PLAYFIELDS

SITES		A	B	C	D
CRITERIA					
SLOPES	0–4%	x	x	x	x
DRAINAGE	Good				x
SOIL	Fair to Good	x		x	x
VEGETATION	Poor to Absent	x			x
SCORE		3	1	2	4

Figure 6-14
Typical spreadsheet for site selection of playfields.

ning of the park. In its original form, the overlay process made famous by Ian McHarg's seminal work "Design with Nature" utilized actual sheets of clear plastic with applied data shown as areas of color or texture, which were then assembled in registration on top of a base map. All the data collected were thus directly related visually to each other on the base map. Computer graphics enable the same process to be performed electronically by digital overlaying, with the powerful capacity of being edited easily and presented in any combination.

Assuming that a computer system will be a necessity for any agency, a conscientious effort toward its care and feeding will include stuffing its data files with statistics and facts about sites, uses, and local population, and with management data and economic information, regardless of whether the specific data are needed at the moment. The stored resource, created at relative leisure, will prove invaluable when needed under pressure.

In the capacity of a user, be wary when evaluating an analysis presented in support of a proposed plan, no matter whether the source is in-house staff or a remote consultant, whether the plan is computer generated or laboriously hand drawn. Sheer volume of paper does not necessarily equal impact. A successful analysis is not measured in weight, only in quality. A large amount of inventoried data is useful only if actually relevant to the analysis performed. How are the *conclusions presented* related to the *data collected*? Why were the data needed? The analysis should be a record of an orderly thought process leading to a logical plan.

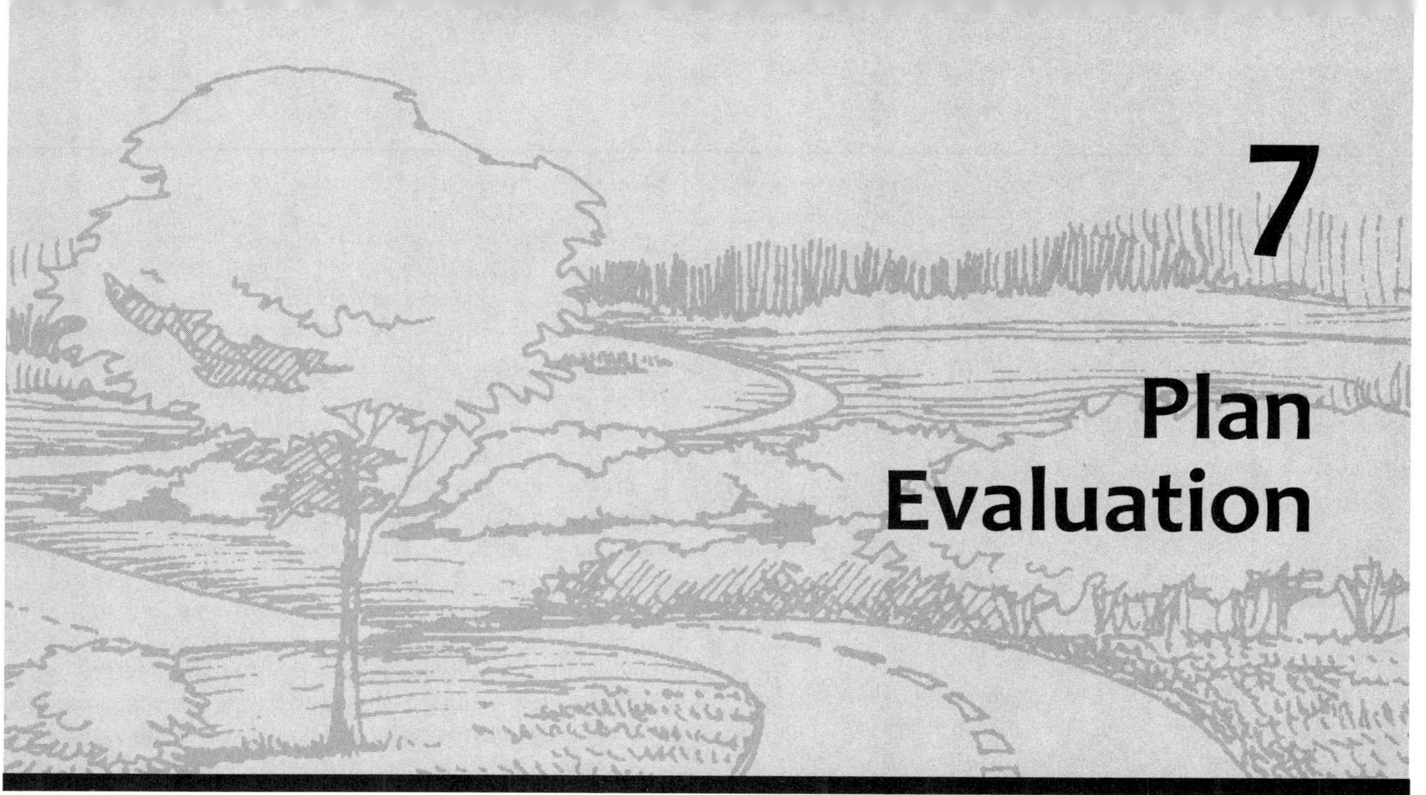

7 Plan Evaluation

Although their chores are different throughout, designers and critics are confronted with many similar situations. Beginning with little understanding of solution direction, both are vulnerable to panic in the face of the project's magnitude and both need a clear view of where to begin. While both might have early feelings about what is needed, such general feelings must be turned into specific requirements in order to direct attention to the significant determinants of the case. The success of the end product—the architect's drawing or the critic's evaluation—hinges on the ability to make correct value judgments about the synthesis.

The best aid that can be suggested for the critic—as well as for the designer—is a systematic approach to sorting out determinants, although this cannot guarantee results since a favorable outcome depends on how well the critic operates within the framework provided by the system. While any evaluation procedure is therefore but a tool, the one suggested here should arm the critics with enough understanding to give them some confidence in their conclusions.

Selecting the Designer

The process of ensuring an appropriate solution to your land-use problem begins long before the plan is developed. It starts with the selection of the landscape architect. If you find a good one, your critique becomes but a cordial review, lasting just long enough to satisfy your curiosity, leaving ample time in the evening to toast both your fine judgment and the landscape architect's at a parlor of good cheer.

When designers are employees of your agency, you already know something about their work and abilities. However, many park departments do not budget a full-time design staff, preferring to contract the preparation of each development plan to a private consultant. Selecting a qualified consultant is an art that should be mastered by the financing party to the contract. It involves a searching analysis of the designer's references and past performances. Several candidates should be asked to present their credentials; while one might appear adequate to the task, comparison with the qualifications of others will confirm a decision that the best has been chosen.

There is frequently a tendency to keep the money at home by selecting a designer from the immediate area, but this should be tempered by a desire to choose a consultant who is detached from the pressures of local politics and vested interests. Whether from around the area or from out of town, the landscape architect must be able to provide an objective solution to the problem.

Those landscape architecture firms that devote all or portions of their practices to the design of recreation areas should be asked to supply examples of prior works similar to the one you are planning. Inspect the graphics and accompanying texts and draw conclusions about depth of analysis and breadth of thought. (Previous chapters have suggested what should be expected of a competent designer.) During the formal interview, quiz the designer regarding the handling of issues that have been discussed here.

While specialty experience deserves consideration, don't overlook landscape architects who have not done park projects, remembering that their education proposes to equip them to treat all types of land-use problems. Even the office with the highest number of recreation areas under its belt started somewhere. In reviewing samples of their work, look for design aspects similar to those associated with parks, such as circulation, use-area relationships, resource utilization, and aesthetic experience. Excellence in these categories and professional presence under examination may very well indicate a capacity to meet your specific challenge and provide the quality you seek.

Under no circumstances let your judgment be swayed by *products* rather than *process*. Don't look at a beautiful portfolio or brochure offered by a park designer and fall in love with the snazzy play equipment selected, the glossy furniture along the walkway, or the brightly colored water spray doo-dads with the squealing kids running through. Look deeper to find the designer who knows how to organize and form the park that will contain those excellent furniture elements. Procedure first; products follow.

Having now introduced the critical process of obtaining the right consultant, recognize that it is a complicated labyrinth, navigated carefully to ensure the best results for every park. We'll explore this in more depth in Chapter 8.

Preparation

A landscape architect—the type from whom you would not want to buy a used car—once said that he could convince a client to accept anything because, when he presented a proposal, he was the only one prepared. Fortunately, such cavalier attitudes are rare, but the lesson remains: All parties to an issue should be familiar with its details if they wish to guard their interests when the issue comes up for discussion. To those who have a stake in the quality of a park design proposal, preparation for discussion means getting at least a feeling for the problem and some understanding of the factors that could affect its solution.

This can be accomplished through knowledge of program requirements and site characteristics, the former gleaned from cursory research or by virtue of personal acquaintance with program activities; for example, you may be an expert equestrian and thus quite knowledgeable about stable development and bridle path layout. Site information can come from a topographic map or an on-the-ground reconnaissance. From these thoughts can evolve a fistful of factors that you consider important in resolving the problem.

If more thorough preparation fits your mood or the circumstances, you can expand your list of important considerations by setting each of the principles and matters of concern against what you understand about the program and site. When played against the particulars of the problem, the principles and matters of concern act as triggers, dislodging from the back of your mind a host of specific issues which you feel the designer should ponder on the way to solution.

To see how this might work, let's assume that a baseball diamond is required by the program. Baseball is selected because most of you are probably familiar with the game and need do no research in order to participate in this exercise. Go down the following list, asking yourself what each item brings to mind in terms of design decisions that should be made regarding the diamond's function and aesthetic appeal.

1. *Everything must have a purpose.*
    a. Relation of park to surroundings
    b. Relation of use areas to site
    c. Relation of use areas to use areas
    d. Relation of major structures to use areas
    e. Relation of minor structures to minor structures
2. *Design must be for people.*
    a. Balance of impersonal and personal needs
3. *Both function and aesthetics must be satisfied.*
    a. Balance of dollar and human values
4. *Establish a substantial experience.*
    a. Effects of lines, forms, textures, and colors
    b. Effects of dominance
    c. Effects of enclosure
5. *Establish an appropriate experience.*
    a. Suited to personality of place
    b. Suited to personality of user
    c. Suited to personality of function
    d. Suited to scale
6. *Satisfy technical requirements.*
    a. Sizes
    b. Quantities
    c. Orientation to natural forces
    d. Operating needs
7. *Meet needs for lowest possible cost.*
    a. Balance of needs and budget
    b. Use of existing site resources
    c. Provision of appropriate structural materials

d. Provision of appropriate plant materials

e. Attention to details

8. *Provide for supervision ease.*

a. Balance of use freedom and control

b. Circulation

c. Safety

d. Discouraging undesirables

Following are some thoughts that these categories might trigger. But don't look at them until you have first gone through the suggested procedure and set down your own ideas. Your purpose will be to identify areas in which design decisions must be made; the form is inconsequential. They may be stated as directives that should be followed, questions you want answered, or specific concerns whose treatment deserves inspection.

As you proceed, you may draw a blank on some items, but this is to be expected since certain categories will be less applicable than others, with a few totally irrelevant to particular facilities. For example, the category *relation of major structures to use areas* is probably irrelevant to the example we are considering: a baseball diamond in a complex of amateur sport fields. But even this negative conclusion has value in preparing for evaluation; having considered the issue, you have some assurance that an exception has not been overlooked.

In addition, do not be frustrated if repetitions emerge as you tick off the categories. Double-checks have been built into the principles and matters of concern in accordance with the contention that if one phrasing does not dislodge a significant thought, perhaps a synonymous one will. That is, if the fact that a baseball diamond needs a flat surface does not dawn on you when you dwell on *relation of use areas to site*, there is a chance that it will be brought to mind by *use of existing site resources*.

When you have completed this exercise, compare your list with the one below to see how close you come, or if indeed you have uncovered other factors.

Some Factors Affecting the Design of a Baseball Field

- Buffer noise and flight of errant baseballs.
- Flat surface draining to periphery.
- Direct access from parking lots or pedestrian entrances to spectator area.
- Location of drinking fountains, comfort stations, concession stands.
- Maintenance equipment stored near diamond or elsewhere?
- Positioning of bleachers or other viewing accommodations.
- Static space.
- Cleared field.
- Dark outfield background.
- Layout dimensions according to rules of the game, including at least 300 feet down each foul line.
- Is fee control necessary?
- Sun out of the eyes of the batter and pitcher.
- Soil appropriate for grass growth.
- Permeable soil.

- Avoid access to other park units across outfield.
- No hidden obstacles in playing field.
- Lighting system on egress routes for twilight games.

Go through this process for all units in the program. While it may take a bit of time to evolve lists of factors for the first few units, the pace should quicken with subsequent ones as your concentration improves and as you begin to pick up common factors; for example, areas for lawn bowling and tennis need to be laid out according to the official dimensions of the games, just as the baseball diamond does. Checking off the principles and matters of concern against all units in the program should also make you conscious of relationships—lawn bowling and tennis courts should be located away from the baseball field—which should be jotted down as they occur to you.

Not much more than an hour of uninterrupted concentration should be all you'll need to effectively run through most programs regardless of their lengths. It is not necessary to work out solutions to the problems you uncover or do more than speculate about how determinants might best be handled. This is what the designer is doing for you. Remember also that, while you may see some directions that you think a solution might take, your ideas are most likely fragments not tied to other considerations. Therefore, you should not carry into critique prejudgments regarding what the final form and organization might be, for your thoughts have not been tested against the full range of priorities. Nor is it required that you come up with an all-inclusive list of concerns. In fact, it would be as impossible to derive an all-inclusive analysis of the problem at this stage as it would be for the designer to predict all aspects of the solution while in the middle of analysis.

With these cautions in mind, any information you gain, no matter how cursory or brief your examination, serves the primary purpose of the procedure: orienting you to the problem so that you will not have to approach the plan drawing cold. By raising awareness of what may be involved, the procedure also sharpens your ability to react to the proposal as presented by the designer. In addition, you have developed a checklist comprising focus points or items that will demand your attention during critique. They can also be starting points from which an ensuing discussion, allowed to take its course, may direct eyes to other significant matters.

A desire to become oriented and identify specific focus points should also motivate your inspection of the site. Warming up for evaluation, you should work toward discovering as many site limitations and potentials as possible given the time allowed for the study and the information you have available.

Your site review may be assisted by the principles and matters of concern, triggering thoughts in a fashion similar to that suggested to help cull information about the program. Or your attention might be directed to site determinants by points on the program checklist. Consider our baseball example. The fact that the facility requires a flat surface and permeable soil should send you searching after those characteristics on the site. Your findings are additional focus points, in this case site potentials and limitations whose management in the solution should be observed. These may be compiled as an addendum to the program checklist or, better yet, jotted down in place on a topographic map of the site. Since the designer will be presenting conclusions on a map, this should facilitate comparisons.

You should now be quite able to approach the proposed solution *on its own terms.*

Critique

The opportunity to evaluate may occur in many ways. It may come in the form of a plan routed across your desk. In this case, the plan is unlikely to be accompanied by its designer. Or you may be given a written report outlining the essentials of the proposal to which the plan is attached. You may also attend a preliminary plan presentation made personally by the landscape architect. While the following approach supposes that the designer is present, most of the steps fit the other instances as well.

I. *Understand what the designer has done*. During the presentation, make notes when questions come to mind. During the discussion period that usually follows the unveiling, pin down the designer for justification. If you cannot follow the reasoning, ask for clarification. If you cannot read topographic relief, character of lines, forms, textures, and colors, and the nature of the spatial sequences from the plan, ask to have them explained. If you feel that something has not been made clear, ask to have it re-explained. If you do not understand the significance of an issue ask. Ask. Ask. There are no stupid questions. Only questions.

II. *Consider the designer's goals*. These are the primary objectives or criteria that support the major design decisions and should show up in the designer's opening remarks or appear early in the written report. Are they valid?

A. If you approve of the purposes, proceed to the next step.

B. If you are not entirely convinced by the reasoning underlying some goals, go on to the next step, but qualify subsequent judgments accordingly.

C. If you feel that the designer's criteria are completely unfounded—they are as muddleheaded as "tree cover should be obliterated inasmuch as it adds nothing to the value of the site"—forget about the next step. Critique only the goals and write off the work as a completely misguided effort.

III. *Evaluate goal realization*. The following guidelines should help you concentrate and also help you sum up your thoughts during presentation and discussion. It can also be used as a procedure for treating in detail those plans that are available for lengthy study.

A. From the plan, abstract its concept. Do this in your mind or on tracing paper and you will have an uncluttered view of the proposal's functional and aesthetic skeleton: its major use areas, circulation patterns, and spatial structure.

B. Either mentally or physically, overlay this skeleton on a site analysis (perhaps the topographic map on which you compiled preparatory impressions). Actually, a concept abstraction and site analysis should be available to you from the designer as part of the presentation package.

1. On these drawings, bring to bear the full force of your knowledge about design in general and the specific thoughts you have regarding the particular project, treating first the *complex at large*. From the schematics, with an occasional allusion to the original drawing, make broad judgments regarding:

a. Satisfaction of stated goals and additional criteria you feel are warranted.

b. Relation of park to surroundings.

c. Relation of use areas to site.

d. Relation of use areas to use areas.

e. Relation of major structures to use areas.

f. Circulation.

g. Spatial experiences.

h. Aesthetic character.

i. Provisions for order and variety.

2. Now go back to the parent drawing and make more detailed judgments about *each major use area.* Go through the areas in some reasonable order: right to left; north to south; as found along the major arteries; passive areas, then active ones; or from day-use units to overnight facilities.

   a. Concentrate on issues a–i, above.

   b. Consider relevant focus points from your checklist.

3. Draw conclusions about *each object in each use area.* Throughout, but especially here, you should consider the degree of design commitment possible at the plan scale and the type of drawing being studied. As we said in the chapter on plan interpretation, the location of light fixtures is not a decision reached at 1 inch = 200 feet; the naming of plant species is not a requisite of the master plan but belongs on the planting plan.

4. At this stage, you should have a comfortable understanding of the solution proposal. Finally, go over all the principles and matters of concern in order to see if any issues remain. Now you are ready to sum up.

Case Studies

Let's try out the procedure. Following are some plans waiting to test your insights. Case 1 (Figure 7-7 on p. 123) is adapted from an actual professional proposal originally drawn at 1 inch = 50 feet scale. The accompanying program (Figure 7-1 on p. 118), site analysis (Figure 7-2 on p. 119), 3-D site views (Figures 7-3 and 7-4 on p. 120), design concept—solution 1 (Figure 7-5 on p. 121), the site plan—solution 1 (Figure 7-6 on p. 122), and summary explanation of the proposal's strong and weak points should make the evaluation easy for you. The summary gives reasons for the conclusions and offers further tips on how to approach the drawings. Your critique, addressed only to the substance of the plan, need not be as verbose. With this in mind, you may wish to attempt a review of your own before turning to the evaluation of Case 1 and comparing your findings with it. With this model view and the next, Figure 7-4, try to see the elements of the program in their positions in Solution 1 as you make evaluations. Walk the site in your mind.

Case 2 (Figure 7-7 on p. 123) is another solution for the same problem. This was not the consultant's answer; it was worked up especially for this book to illustrate some errors that could render a

A SCHOOL PARK PROGRAM

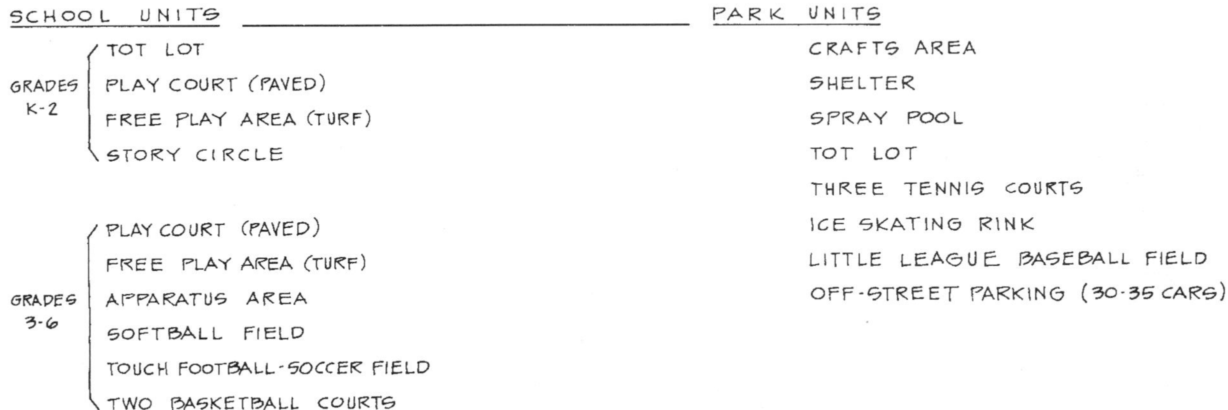

CRITERIA

1. ISOLATE SCHOOL FACILITIES FROM PARK UNITS TO MINIMIZE OVERLAP INTERFERENCE DURING SCHOOL HOURS. YET, IN ORDER TO TAKE ADVANTAGE OF THE SCHOOL-PARK CONCEPT (AVOIDS FACILITY DUPLICATION; REQUIRES LESS LAND THAN COMPLETELY SEPARATE DEVELOPMENTS), ACCOMMODATE A DUAL USE FLOW FOR THOSE PERIODS WHEN SCHOOL IS NOT IN SESSION.
2. PROVIDE EDUCATIONAL EXPERIENCES.
3. RELOCATE EXISTING SERVICE DRIVE TO MINIMIZE HAZARDS OF MIXING KIDS WITH VEHICLES.
4. PROVIDE RELIEF FROM THE LOCAL ENVIRONMENTAL DULLNESS THAT RESULTS FROM THE COMMUNITY BEING SURROUNDED BY ENDLESS FLAT CORNFIELDS.

Figure 7-1
A school park program.

proposal useless. No explanation is provided; you will have to find the weaknesses yourself, using the same program and site analysis, but abstracting your own version of the concept.

You are on your own for the four subsequent cases (Cases 3–6); here you will evaluate alternate design proposals for two new parks on undeveloped sites. Following the new-park studies are two redevelopment studies (Cases 7 and 8) for existing parks. We will further discuss the concepts of life cost and obsolescence in connection with these case studies. Case 9 is a composite study of a variety of the design and development of linear parks. Cases 3–9 are all hypothetical, created for this book to illustrate a number of the points we have discussed. They typify many situations you might come up against when dealing with "live" circumstances. In all cases, you must supply the concept abstraction and evaluation.

Both problems and solutions for the new-park studies are presented in simplified form on the assumption that these are among your first attempts to evaluate a design proposal. You should also be aware that a complete examination of program and site requires more knowledge than you can glean from the program and map provided. For instance, little is indicated about surrounding influences and there is no description of applicable ordinances. Unavailable too are the

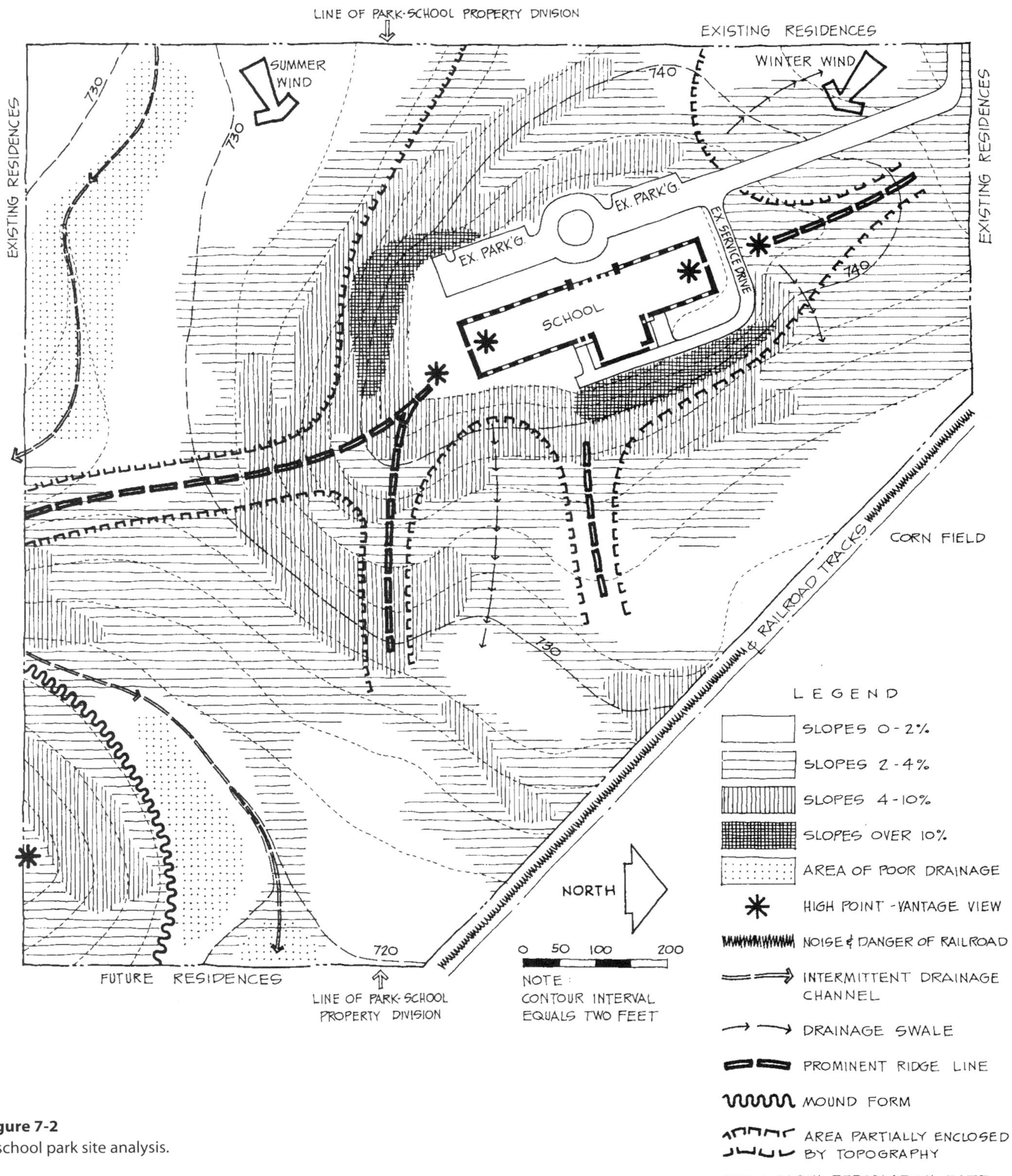

Figure 7-2
A school park site analysis.

Figure 7-3
Topo model of site, viewed from East toward West.

Figure 7-4
Topo model of site, viewed from Northwest toward Southeast.

nuances perceived by someone who has walked the site and is familiar with the locality and the affected population. While this knowledge would be handy in actual circumstances, enough information is presented in these hypothetical cases to confront you with a meaningful challenge.

Until you become quite well acquainted with professionally prepared plans and their ingredients, it is suggested that in evaluating these cases you attempt to follow the procedure outlined in this chapter from preparation to critique. This is not to promote the procedure as the only one possible, nor is it to imply that everyone will feel comfortable using it. Indeed, there are circumstances where it will prove to be unwieldy, but since it is comprehensive and offers a technique for unearthing specific determinants, its use at the onset should encourage the development of similar traits in whatever other system is eventually embraced. While you are still a neophyte it should give order and fullness to the review; later it can serve as a plane of departure from which a personal method for scrutinizing the anatomy of a park plan may be developed. Such a pattern is sure to emerge as your maneuvers become progressively more automatic.

Case 1: Evaluation of a School Park Site Plan

Note: The solution evaluated here was adapted from a plan for the Thomas Paine School-Park, Urbana, Illinois, prepared for the Urbana School District and the Urbana Park District by landscape architect Phillip E. DeTurk.

Goal Validity

1. The decision to separate school units from park areas seems appropriate. To minimize the chaos that could be caused by competition for facilities, an organizational system must be developed that will allow vacationing adults, tots and parents, adults on afternoon breaks, and others who might be in the park during school hours to have free run of the areas most attractive to them. Yet, their noise and physical presence must be kept from disturbing school kids and teachers. This separation should not be difficult to achieve. The trick will be to come up with a plan that will both achieve this goal and satisfy the seemingly contradictory demand: fusing school and park facilities so that after-hours and summer programs can be efficiently administered over the entire site.

   A qualification is in order. Essential to the success of any school park operation is a prearranged agreement between school and park authorities regarding responsibility for staffing, maintenance, and liability for injury when one authority programs activities on another agency's land. Unless this is spelled out and followed through, the concept is in trouble no matter how well the facilities are organized.

2. Arguing against the goal of providing educational experiences is like disparaging motherhood and apple pie, especially in a development associated with a school.
3. Foresight would have eliminated the need for the financially wasteful third criterion. For, while traffic on the service road is

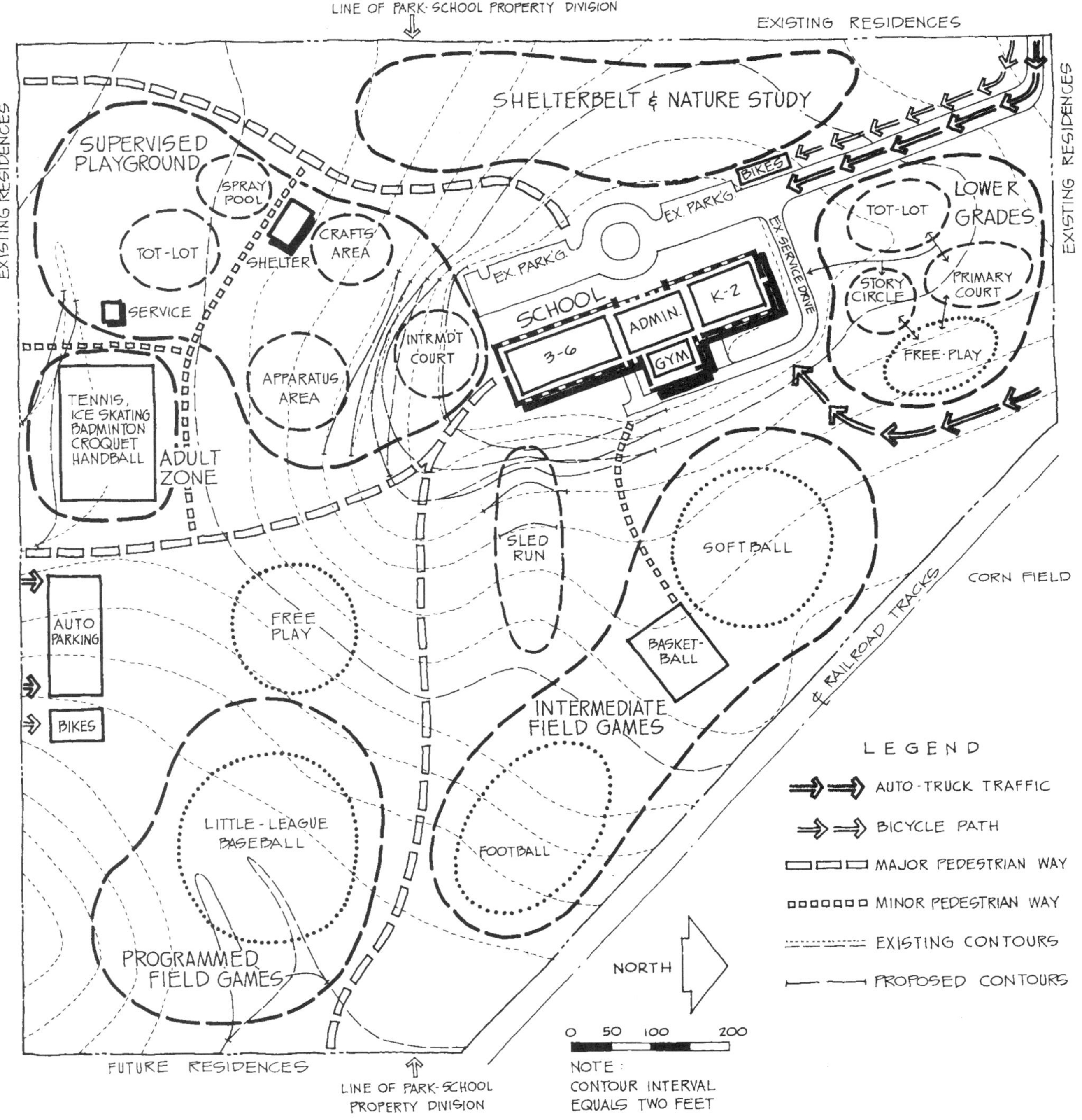

Figure 7-5
A school park design concept—solution 1.

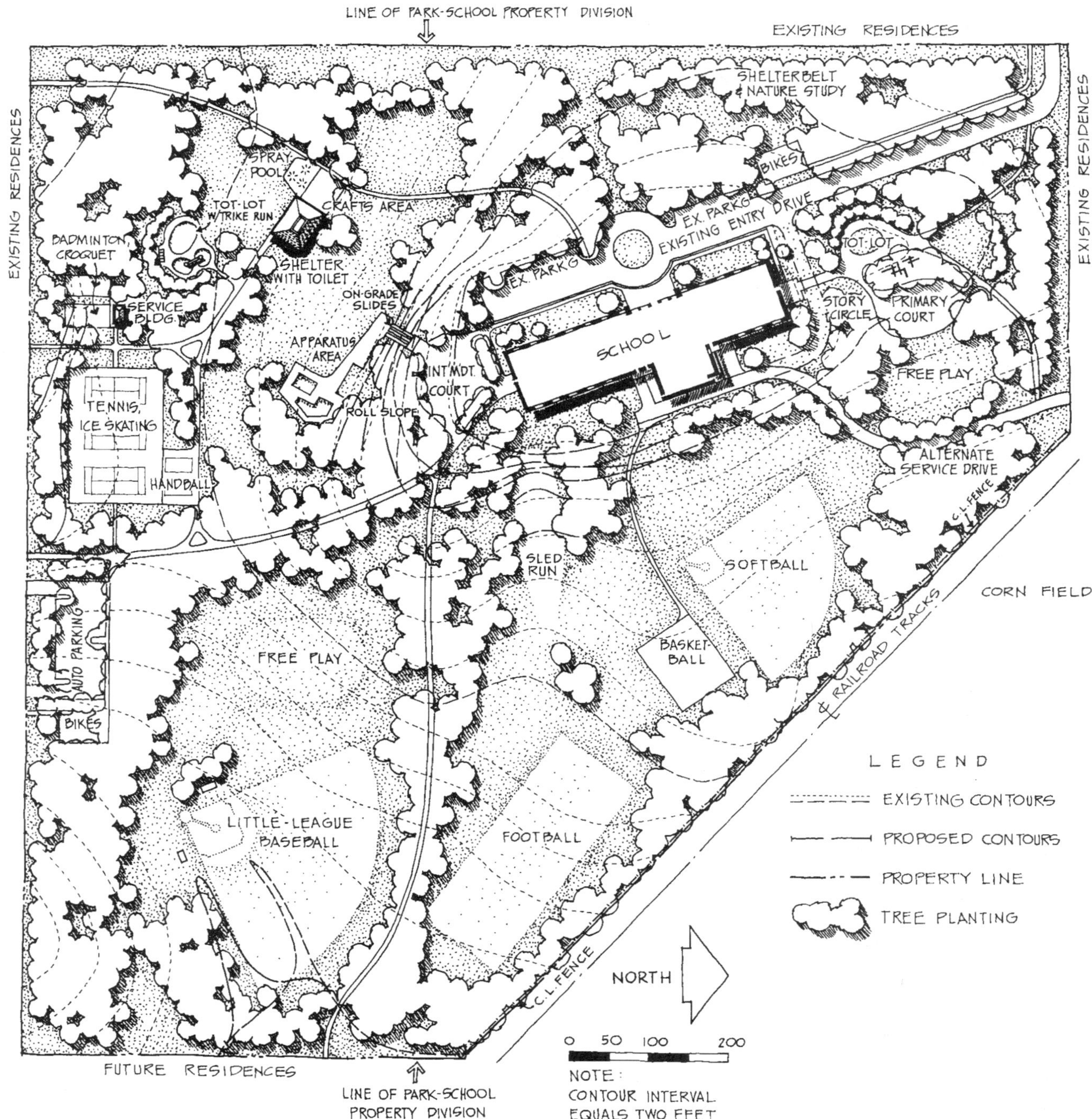

Figure 7-6
A school park site plan—solution 1.

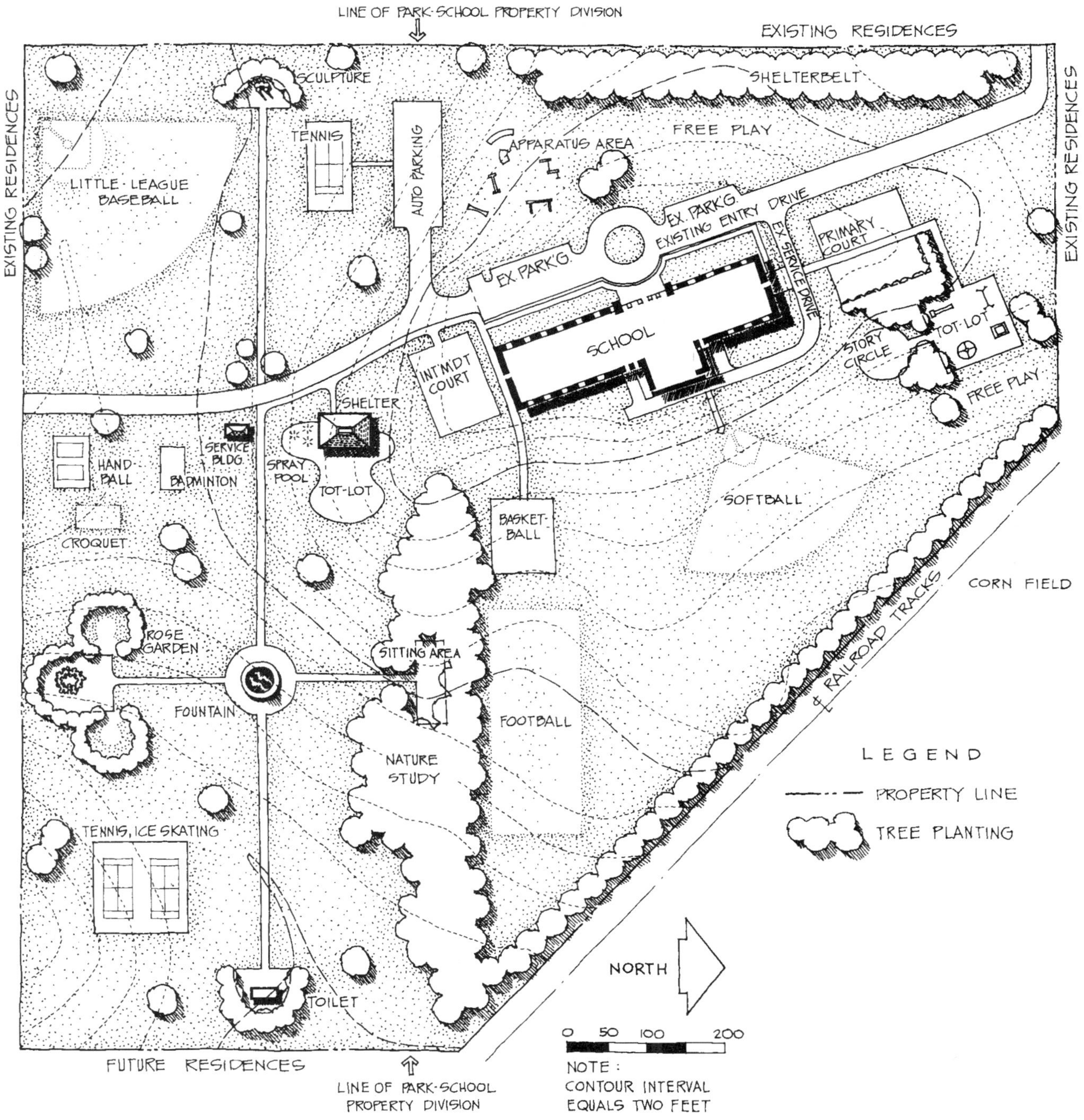

Figure 7-7
A school park site plan—solution 2.

only sporadic, hence posing but a moderate hazard, the fact that the road was built where it is at all illustrates in microcosm what can happen when building design and site planning are not coordinated.

The school structure was sited and built before the landscape architect was approached to consider the rest of the land. When the landscape architect is not present to consult with the building architect, an overview of the entire development is missing. As a result, the building architect, unable to judge the effects of such decisions on land-use efficiency, may very well divide up land suited for companion use units by access roads or locate buildings on land more appropriate for other purposes, to give only a couple of examples. This also works in reverse. When laboring in isolation, the landscape architect may tie down a building to a place on the site that could pose restrictions on the design of the structure. What should have been called for, as it should be in any project comprising buildings and extensive land usage, was a collaborative study of the entire 33 acres by building architect and landscape architect, treating the full acreage as a single problem.

However, hindsight will not rectify this mistake. Since the service road abuts the building at the door out of which the kindergarten kids are expected to flood, and it can be anticipated that this age group will be oblivious to the potential danger, the road should be relocated. Although traffic is sporadic, the prospect of but one accident is too chilling an alternative to leave it alone.

4. Those who have experienced central Illinois know it as a flat sheet of corn and soybean fields that presents itself to the eye in expanses guaranteed to anaesthetize the senses. Too much of a muchness, as one wag has put it. To secure environmental diversity is therefore a worthy criterion. How this is done deserves special attention because, even though the basic charge is to instill variety, the character of the place must also be retained in order to avoid the embarrassment of establishing a 33-acre sore thumb.

The Complex at Large

Goal Realization

Separation is achieved simply by placing most of the school units on school property and the majority of park facilities on the park-owned land. No big deal. However, the designer receives a standing ovation for the balancing act he has done with the use areas within each property. Note first of all that the tennis and handball courts and the "tot lot" and park shelter, which are high-attraction units for school-hour park visitors, are placed at some distance from the school building, which also puts them handy to the park entrance. Thereby, the park population is concentrated where conflict is unlikely to occur during school hours. In addition, the park's play apparatus and free-play space, which may have minor use during the school day but be major attractions after classes let out, are separated from the building by sloping topography. Note, though, that they are adjacent to the school's intermediate court and are thus available if school leaders desire. Now, refocus on this quadrant. You can see a complete activity setup suited to an after-hours or summer program for young children, with the shelter serving as a centrally located supervisory station.

By virtue of its proximity to both the school's intermediate court and game fields, the free-play area can be used in conjunction with the former and as overflow for the latter, thus creating use possibilities for the park's open area, an area which might otherwise lie fallow during school hours. Flowing together, yet with delicate separations preserved, the full complement of school-park fields offers a range of combinations for evening, weekend, or summer sport programs—big areas, small areas, places for structured events such as playground tournaments, spaces for unprogrammed activities such as kite flying, model airplane soaring, or just plain running around. Between scheduled events, the field half of the site is an ideal arena for satisfying whim—from pickup games to unpredictable role playing, up and down its slopes, under and around its trees, in its small and large spaces—for it is well buffered from the game courts and the youngest children's activity areas where greater use control must be exercised.

A break in this pattern may be noted; the basketball court is located near the softball field rather than proximate to the tennis courts where it would be more available for adult school-hour use. However, since it has been requested as a school-related unit and is likely to be of greatest attraction to the after-school set, a location removed from the other courts seems justified.

Therefore, the layout appears to meet the first goal; it contains enough flexibility to fit a range of use combinations yet has an appropriate number of relationship safeguards to ward off use conflicts. There is only one major duplication, that being the "tot lots," which show up both north and south. This appears unavoidable since tot lots may be demanded by both residents and kindergarteners during the same hours, and, with the building already in place, there is no way a single area can meet these requirements. Given this double demand, it is as reasonable to place the school's lot next to related classrooms as it is to site the other near the park entrance. But even here, the designer is on his creative toes, providing the park's lot with a trike run and thereby lending it a special flavor. Coincidentally, a bonus is gained by providing two tot areas. After school hours and during the weekends and summer, residents to the north have a tot lot more convenient to their homes than the one slated for the park proper.

To the possibilities for education that already exist on the site—the nearby railroad, including the original prairie grasses that still grow along its right-of-way, the adjacent cornfields reflecting the region's agricultural heritage—the designer has added a nature study shelterbelt and encouraged human interaction by providing several catalysts for gatherings. Judgments on questions such as the suitability of the crafts facility and the educational horizons of the play apparatus must wait until more detail can be seen; the scale prohibits full disclosure of the designer's intentions for these.

The nature-study area suggestion is an illustration of a liability turned into an advantage. Shortly before the preparation of this plan, the roof of the school was lifted from its rafters by storm winds, and the pieces scattered to Munchkin Land. Fortunately, there were no children in the building at the time. Wind screening is therefore advisable, lest this happen again with more dire results. The designer translates the shelterbelt into a nature-study place, thereby exploiting the dual-use principle. Happily (since the designer had no choice as to its location), the shelterbelt lies adjacent to the school and reaches conveniently toward the park shelter, from which interpretive programs can emanate.

While human interaction can take place wherever there are people, in this plan design ploys have enhanced the possibility. The park shelter, centrally placed for observation of child play, and the spectator slopes, convenient to the softball and little league fields, are two gathering places that are immediately evident. Other potential sitting spaces associated with tot lots and game courts remain to be exploited in further detail studies.

The handling of the service-drive relocation is another example of turning an obstacle into an asset. Once obstructing the passage from the classrooms to the area best suited for primary play, the road now separates the younger children's activities from the older kids' sport fields, thereby discouraging overlap. The drive still nestles next to the gym, but this seems unavoidable because of the location of the building's service docks. While this expanse of pavement at the foot of the building will still look unattractive, the remaining safety problem is relatively negligible. It can be assumed that traffic movement has subsided at this point, the road becoming primarily a storage and maneuvering surface, and that the older children passing across the pavement will be more alert to the vehicles than the primary graders.

When it comes to the fourth goal, instilling environmental variety, the designer is in luck, for the site sits on one of the glacial moraines that occasionally interrupt the pervasive flatness of the region. As fortunate as he might have been to begin with, the manner in which he has followed through must be logged to his personal credit. Recognizing that the site itself supplies a welcome contrast to its surroundings, the designer has intensified its character, thereby managing order and variety with the same stroke. Use units have been settled into existing pockets formed by the site's rolling topography. The sculptural character of the ridges has been accented with brows of tree masses. And to ensure that users experience the views associated with the changes of grade, collector walkways have been placed on the crests of the topographical rolls.

Relation of Park to Surroundings

The shelterbelt screens the school's parking lot and entry road from residences to the west. Reaches of pavement required by tennis and parking to the south are also adequately buffered from residential view. The tot lots are guarded from surrounding traffic by distance and plantings. Areas of highly concentrated use have been kept away from the railroad. The chain-link fence proposed to edge the right-of-way should keep bouncing balls from the tracks, while the high trees should handle errant flies. The fence should also thwart those who would wander onto the roadbed at times when there is no teacher supervision.

Relation of Use Areas to Site

In addition to recognizing the existence of topographic pockets, the designer has perceived that variations in grade are a distinguishing characteristic of the site. In matching uses to the land, he has been faithful to the following realities regarding relative slope severity:

1. Since slopes of 0 to 2 percent are essentially flat and the soil that covers the entire site is heavy and resistant to immediate percolation, drainage would be slow if they were planted in lawn, but rapid if they were paved. Hence, slopes of this type are suited for court games such as basketball, tennis, volleyball, and others.

2. Slopes in the 2 to 4 percent range are fairly flat, yet steep enough to provide adequate surface-water runoff if planted in lawn. These slopes are appropriate for sport fields: softball, baseball, football, soccer, and others.
3. Slopes ranging from 4 to 10 percent have rapid surface runoff but are too steep for organized field sports or court games. If they are planted in lawn and intensive use imposed, erosion could be a major problem. Slopes of this type can be used for general free play, where use is sporadic and does not conform to a set pattern.
4. Slopes over 10 percent are too severe for concentrated use. Erosion is a definite problem requiring such slopes to be stabilized with ground covers, rough-cut lawns, trees, and other soil holders. These grades should receive only intermittent traffic or special use where steep pitches are essential to the play experience (see built-in slides, sled run, roll slope, and other uses on the plan). These areas can also serve to separate incompatible activities.

Only slight earth reshaping is needed to accommodate these appropriate matches. This occurs notably south of the school, extending the existing shelf to a size adequate for intermediate court installation, and near the tennis courts to the southwest and the little league field to the southeast. The latter moves relocate the existing swales in order to improve drainability, while allowing the site's natural drainage pattern to remain intact.

Relation of Use Areas to Use Areas

Most of these have been covered under goal realization. In addition, note that the park's car lot is central to the game courts and little league field, which makes sense inasmuch as these facilities will draw many adults who will arrive by car. Primary adult attractions like tennis, handball, and badminton are well buffered from nearby play areas slated for youngsters.

Relation of Major Structures to Use Areas

While students deserve a better view from the western windows than a parking lot, the landscape architect had no control over the location of the lot and entry drive; their placement was determined when the building was constructed. Since the kindergarten, first, and second grades occupy the north wing of the school, the proximity of the tot lot and primary court is advisable. There is similar wisdom in the proposed location for the intermediate court, adjacent to the door that leads to the third to sixth grades. Commotion associated with these areas is subdued somewhat by plantings. The playfields are handy to the gym entrance, noise being minimized by distance.

Circulation

As has been stated, vehicular access and parking lots for the school were set before the site study was begun, and the relocation of the service drive has already been discussed. The parking lot to the south sits on the park periphery, happily minimizing vehicular penetration of the grounds.

Bicycle routes also penetrate the site only slightly, thereby freeing the bulk of the acreage for unimpeded pedestrian travel. It is proposed that bicycles be stored both near the school and in the park in spaces next to the vehicle parking lots, which seems reasonable although, after alighting, bicyclists must cross the entry drive to get to the school entrance, presumably at the same time that school officials are driving up the road to the building. This bug remains to be worked out. Separate bicycle lanes are wisely suggested to avoid the hazards of mixing pedestrians with wheeled vehicles.

Pedestrian access seems sufficient, with walking routes across the park providing for a direct flow to the school. This is certainly necessary due to the amount of daily traffic seeking that objective. The fact that walks follow the ridgelines wherever possible not only suggests the availability of views, but ensures quick drainage during inclement weather as well. Paved access from the gym to the basketball courts and ball fields also eases the need for follow-up maintenance when the ground is soggy.

The three-prong circuit from the southern and eastern edges to the school additionally serves as collector routes for the park from which secondary walkways extend to various interior facilities. Walks are adroitly placed between use units so as not to interfere with play, for the most part also acting as psychological barriers between separate facilities.

Paved access to field areas is adequate, since entry and exit patterns will be unpredictable and will not be confined to the paved walks.

Since service circulation for park maintenance will be minimal, the walkway system or lawn areas can be used as necessary.

Spatial Experiences

A full assessment of spatial character requires a review of commitments still to be made during subsequent detail studies. However, at this scale it is essential that the designer set up the overall three-dimensional network that will govern follow-up thinking. Therefore, what deserves attention on this plan is evidence of a spatial structure per se. This landscape architect has met the objective well, compartmentalizing use units with tree masses or in topographic pockets reinforced with plant material, closing off one compartment from another where separation is advisable, and linking them where circulation or viewing suggests a need to do so.

A clue to sensitivity to the need for three-dimensional thinking is found in a designer's massing of major plant material (rather than scattering isolated trees about as if they had been thrown at a dart board by a drunk). Where the plant materials are attractively massed on a large-scale plan, there is an excellent chance that detail studies will establish appropriate spatial qualities. However, if spatial relationships are not considered at the larger scales, as exemplified by an inebriated attitude toward the plantings, it is unlikely that they will be well handled in later stages of the project.

Aesthetic Character

A complete evaluation of the proposal's aesthetic qualities must also wait until the design of each use unit can be seen in full detail. However, as with spatial structure, what can be sensed at this scale is aesthetic *potential* as indicated by the general appearance of lines, forms, textures, colors, and spaces. While not final commitments, these show what the designer has in mind and will strive to develop in later studies.

The lines, forms, and spatial network should be given the closest attention. At the scale of this plan, textures and colors are used primarily to separate the various ground surfaces to enhance the drawing's appeal and readability. However, at the same time, they might also reflect something about the project's liveliness or sterility.

Are experiences provided? Will they be substantial? This plan seems to say yes, for it contains not only a strong spatial structure but a consistent form "flavor" as well. *Both* spatial and line configurations are spirited, complementing each other and hinting that the designer has indeed opted for a dominant effect.

Will the effect be appropriate? The answer once more appears to be yes, for the plan's parts have an animated and playful feeling, well in keeping with a recreational enterprise of this type. The organic quality of the spaces and forms also suggests a fidelity to the rolling topography of the site.

Order and Variety

Points already discussed, such as fitting use units into topographic hollows, remaining faithful to slope, and intensifying existing site character, leave little doubt that the blending of new with old will come off successfully, thereby fostering environmental order.

Variety potential shows up in the tree-accented topographic changes, varying sizes and configurations of spatial openings, and the exploitation of views. Additional enrichment should arise from the detailing of the use areas and selection of plant species. While this remains to be done, the fact that the designer has indicated so much feeling for the problem within the restrictions of the plan scale gives a reasonable assurance that his detail work will be equally successful. A less confident posture would have to be struck had the designer been lax in presenting such clues to his aesthetic sensitivity.

Each Use Area in Turn

Some general comments are in order before we proceed on the tour of the use units. Additional design steps are required after acceptance of this plan, and focus on each area may trigger suggestions to be incorporated into ensuing detail studies. It benefits both client and designer to think ahead in order to initiate further study on a basis of mutual agreement. Comments augmenting the plan are usually volunteered by the designer or brought out by the client's question: "What else do you have in mind for this area?" Therefore, ideas for future study are expressed below, as they might be during the course of a typical design presentation. The reader must clearly distinguish these ideas from objections, which are also raised.

The first thing you should notice as you begin the evaluation of the individual units is that use units not originally required by the program are proposed. The additions all seem compatible with project purposes and fit in well with the other areas. The service building and facilities for handball, badminton, croquet, tricycling, and bike parking are adjuncts to initially specified units. The nature-study shelterbelt and sled run are further exploitations of the site. Assuming that the designer has determined that such activities fill a neighborhood need, plaudits are due for appropriately broadening the use of the property.

In addition, note that all areas seem to be sized properly and oriented correctly. An exception in the latter category is the sled run slated for an eastern slope, not an ideal orientation. Such a compromise may be excused, for the run is located on the longest steep surface on the site.

By now, evaluation has been almost completed, for during the complex-at-large investigation much has been said about the functional and aesthetic qualities of each use unit. This round will begin with the school's tot lot, then move clockwise, stopping at each area to clean up what might have been missed previously.

Interior circulation works well north of the building, for the route to the primary court bypasses the tot lot, thereby minimizing interruptions. The free-play turf is well located for handling overflow from the court. The entire cluster can be readily supervised from one spot. Also easing supervision, the tot lot is well contained by plantings as is the free-play space, the latter safely barricaded from the service road. In his detail studies, it is hoped that the designer will retain the flowing forms and provide rich material contrasts in order to stimulate imaginations and sensory faculties. It is also hoped that equipment design will provide for a full range of play experiences: sliding, climbing, running, jumping, rolling, balancing, and digging. In this rather small space, this will probably mean the inclusion of several multipurpose pieces; the square footage does not appear adequate for many separate items.

The basketball court will have to be buffered from the softball diamond's foul line. Visual supervision of the entire field-sports area can take place from the walkway. In the actual placement of trees on the sledding slope, the need to maintain unobstructed sliding channels must be considered. Utilizing the vacant ball fields for the sled landing exemplifies the dual-use principle.

Affording a view of the little league field, the nearby mound should attract spectators, thereby encouraging the experiencing of the highest point in the park. The adjacent free-play space can be used for practice and little league tryouts, establishing another possibility for dual use.

Dual use also shows up in the designer's thinking in the employment of the tennis courts for ice skating. Surfacing material that can withstand the rigors of freezing and thawing will have to be selected to make this idea work. Expansion possibilities for the tennis courts seem to be thwarted by their being hemmed in on all sides by other construction. Tennis demand is frequently underestimated, and if a clamor for more courts should be raised, a real problem exists. To make the courts available for those whose schedule permits only evening play, night lighting should be considered in the detail thinking, as should sitting accommodations for those crowded days when players must wait their turns. Since it is centrally located among the tennis, badminton, and croquet areas, the service house can be used to control reservations and equipment loans.

The tricycle circuit suggested for the tot lot can provide a host of fun possibilities if eventually laid out with varying curves, grade changes, and tunnels. Some precautions will have to be taken to minimize conflict with whatever more-sedentary pursuits are planned for the same area. As with the north lot, seating accommodations should be slated for the comfort of supervising parents.

The play spaces surrounding the park shelter are clustered well, allowing sweeping visual inspection from the shelter at the hub of the complex. The spray pool (a safety problem) and crafts area (requiring direct student-teacher contact) are wisely made to abut the shelter. However, measures will have to be taken to keep wind-whipped spray in check. Further exploiting dual-use possibilities, the crafts surface and structure may serve for small group picnics held as part of a summer playground program. Consideration of this possibility leads to the thought that the shelter might contain equipment storage facilities and an outdoor fireplace but raises a question about the inclusion of toilets. While a convenient comfort station is desirable, the flushing sounds sure to be heard through the walls might very well dull appetites. Perhaps public toilets could be incorporated into the service building a few yards south, thereby providing a handy convenience for adults as well.

Situated well on the other side of the shelter from the tot lot in order to minimize age-group conflict, the on-grade slides, apparatus, and roll slope present intriguing possibilities for stimulating play. The plan sets this up as an action area, a theme that should be carried through in the detailing. Accordingly, criteria for selecting apparatus should include the pieces' potential for stimulating adventures, fostering role playing, and triggering imaginations. Grass chosen for adjacent slopes should be the toughest strains in anticipation of the activity that will be encouraged upon them.

To fully exploit the potential of the shelterbelt as a nature area, possibilities should be left open not only for plant identification but for insect, bird, and small wildlife study as well. The grove should therefore be allowed to mature in its own way, with undergrowth taking over naturally. This rules out grass mowing. Undoubtedly, to get the shelterbelt going, the installation of small trees will be called for, since the cost of covering such an expanse with mature specimens would be prohibitive. Accordingly, if the plan is to be implemented in phases, the belt should be included in the first stage. Then it can be growing to full usefulness while funds for the remaining work are being pursued.

Objects within Each Use Area

Other than the few speculations above, no questions can be directed to this plan regarding such individual items as drinking fountains, curbs, drainage catch basins, or trash receptacles. Critical commentary on objects in the use areas must wait until commitments at detail scale are seen.

Summary

Distinguished by balanced attention to aesthetics and function, order and variety, spatial and ground patterns, use freedom and control, people and mechanical devices—and by an organizational system tailored to the site that separates school from park yet produces an overlap appropriately patterned for maximum use flexibility—this is an excellent plan. Most concerns raised during the critique involve matters easily handled in the detail stage without affecting the major components of the proposal. The only real problem foreseen is the difficulty of expanding the tennis courts. If they have a handle on demand, park authorities can readily judge the degree of risk involved. Hence, the decision to revise the court area or leave it as proposed rests in their hands.

Case 2: Alternate Solution for the School-Park Site

Critique the alternate solution in Figure 7-7 (on p. 123). What aspects of this plan make it inconvenient for users and inappropriate to the site? What favorable comments could you make, if any? Are there any aspects of this proposal that you think could be incorporated into the DeTurk plan? What significant difference between the two plans should immediately clue you in to the inadequacy of the second proposal? Go through the critical procedure as in Case 1, then compare your results with the results for Case 1.

Comparison of Priorities in Case Study Evaluations

Before proceeding to the following case studies, it is important to recognize the significance of relative size or scale of the various park areas. Look at the size comparison diagram (Figure 7-8) to see the relationship in size of each of the properties being considered. The purpose of this diagram is to offer a guide to the level of detail that can be considered at each scale. The scales discussed in chapter 5 under "Plan Types" will help again in this evaluation. For example, the "State Park" project solutions are vastly larger than the "School Park" sites just preceding. Consequently, an evaluation at the level of detail devoted to the School Park should be completely unrealistic for the purpose of this kind of exercise. However, the "Urban Parklet" is

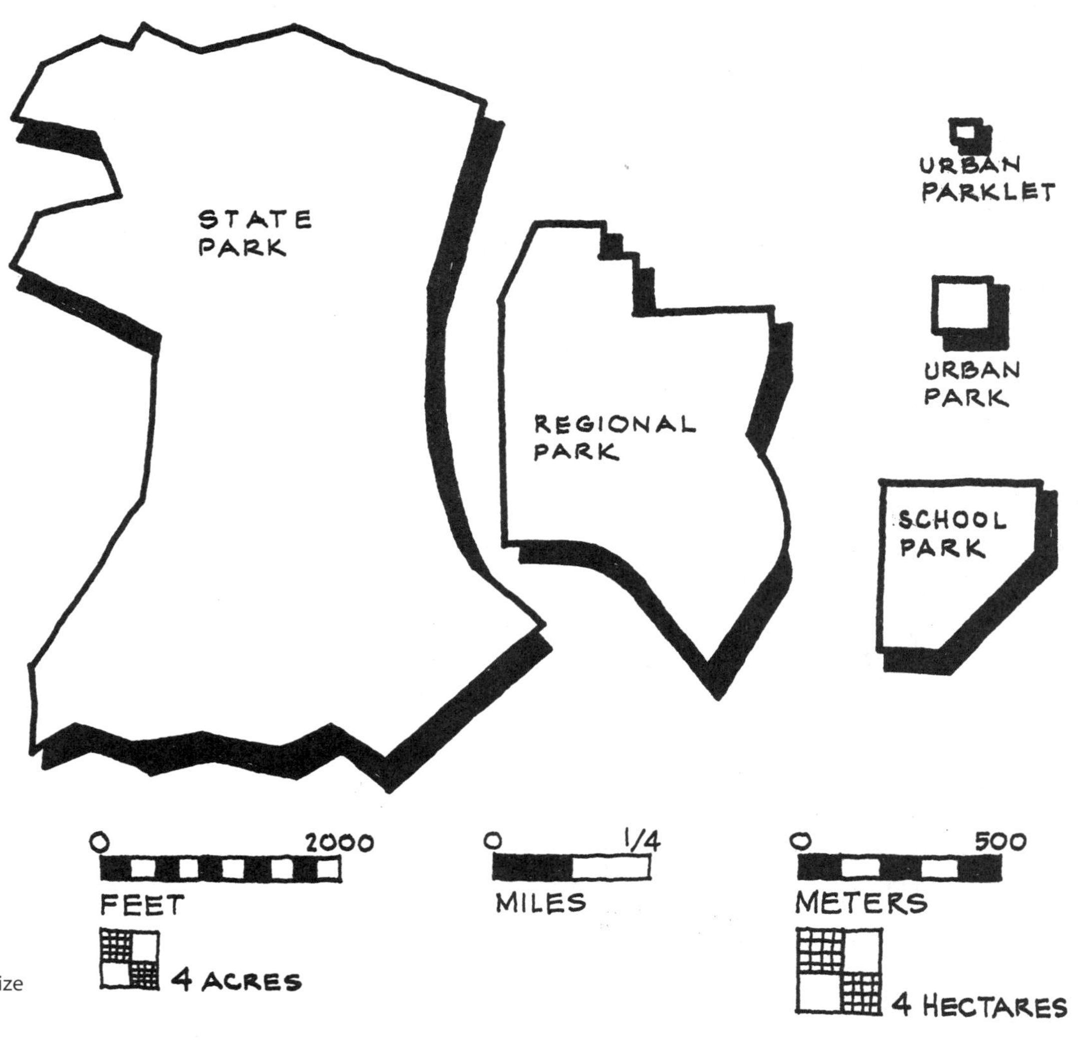

Figure 7-8
Case study sites size comparison.

far smaller than the School Park and so can appropriately be studied in great detail. More notes to this point accompany each of the case study sites.

Cases 3 and 4: Alternate Design Proposals for a State Park

Cases 3 and 4 are alternate design solutions for another new park: in this instance, a state park on an undeveloped site. You are provided with a program (Figure 7-9) comprising a list of uses and an initial criterion, or goal statement. There is also a topographical map (Figure 7-10 on p. 134) with some notes on soil types and slope limitations (assume that these have been transferred from other documents that have not been made available to you) and a site analysis (Figure 7-11 on p. 135). Evaluate the two solutions (Figure 7-12 on p. 136 and Figure 7-13 on p. 137), beginning by commenting on the initial program statement. The scale of the original plan drawings is 1 inch = 100 feet.

A STATE PARK PROGRAM

UNITS

LODGE BUILDING -
GUEST AND MEETING ROOMS, RESTAURANT, RESERVATION DESK FOR LODGE AND CABINS.

LODGE PARKING -
30 SPACES FOR GUESTS, 50 SPACES FOR GENERAL USE, 10 SPACES FOR EMPLOYEES.

30 HOUSEKEEPING CABINS

CABIN PARKING -
ONE SPACE EACH CONVENIENT TO EACH CABIN.

CAMPING AREA -
200 SITES

FAMILY PICNICKING AREA -
100 SITES

OUTDOOR AMPHITHEATER

BOAT DOCK -
ROWBOATS, CANOES

WALKING TRAILS

RENTAL STABLE

BRIDLE TRAILS

MAINTENANCE BUILDING AND SERVICE YARD

STAFF HOUSING -
DETACHED HOUSES FOR MANAGER AND ASSISTANT MANAGER AND FAMILIES.

CRITERION

PROVIDE OPPORTUNITIES TO EXPERIENCE THE NATURAL QUALITIES OF THE SITE.

Figure 7-9
A state park program.

A STATE PARK SITE ANALYSIS 1 OF 2

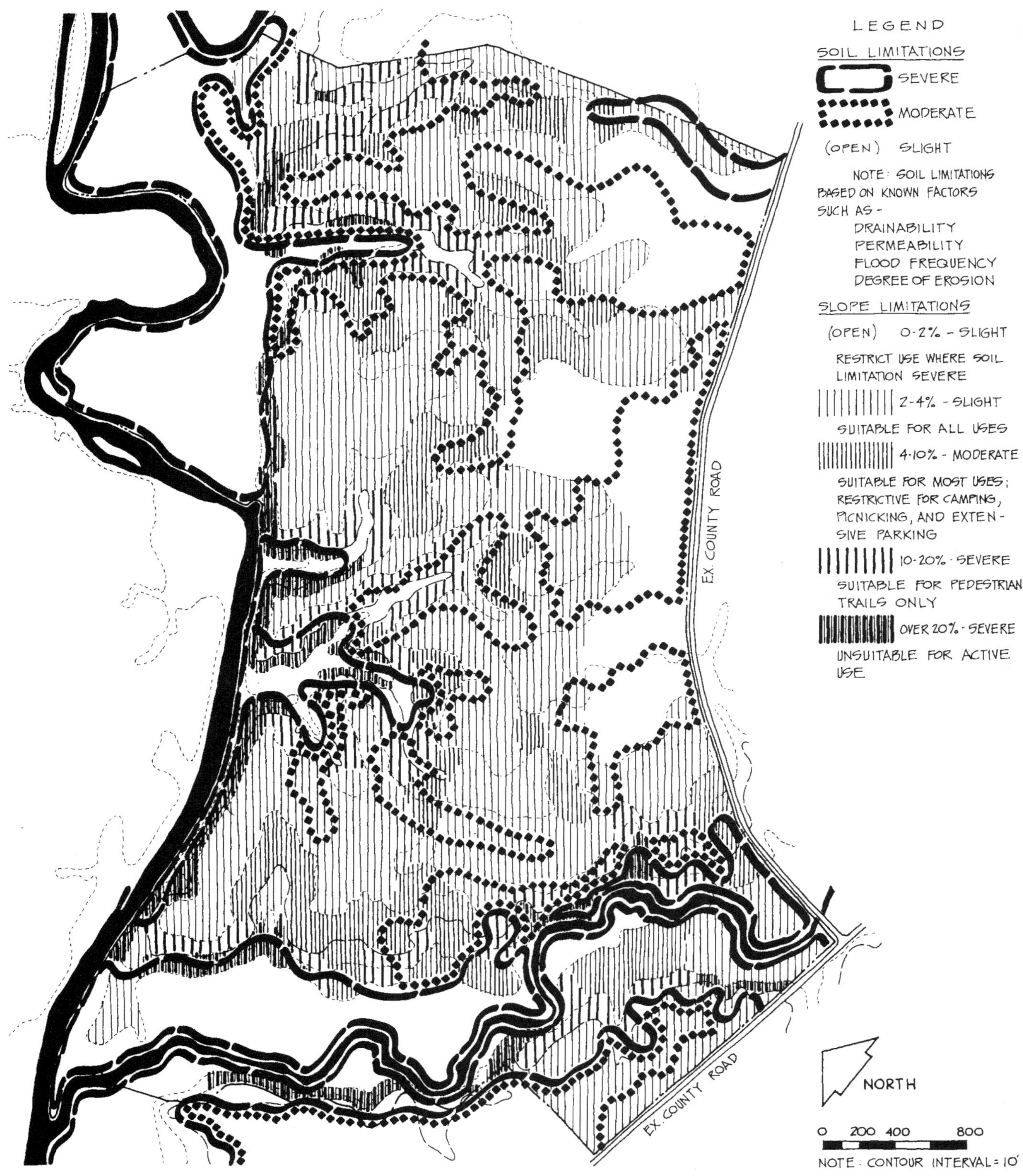

Figure 7-10
A state park site analysis, 1 of 2.

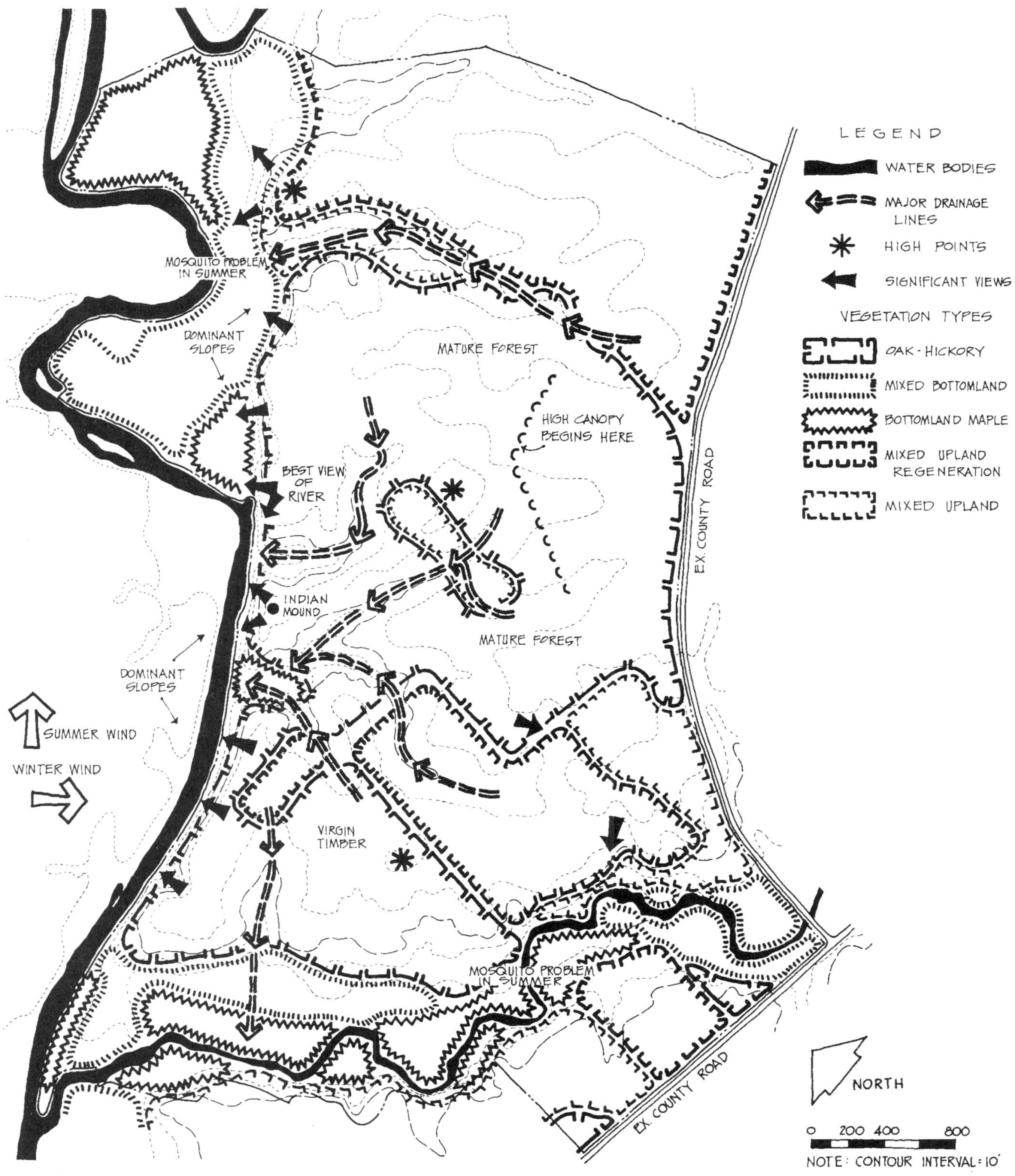

Figure 7-11
A state park site analysis, 2 of 2.

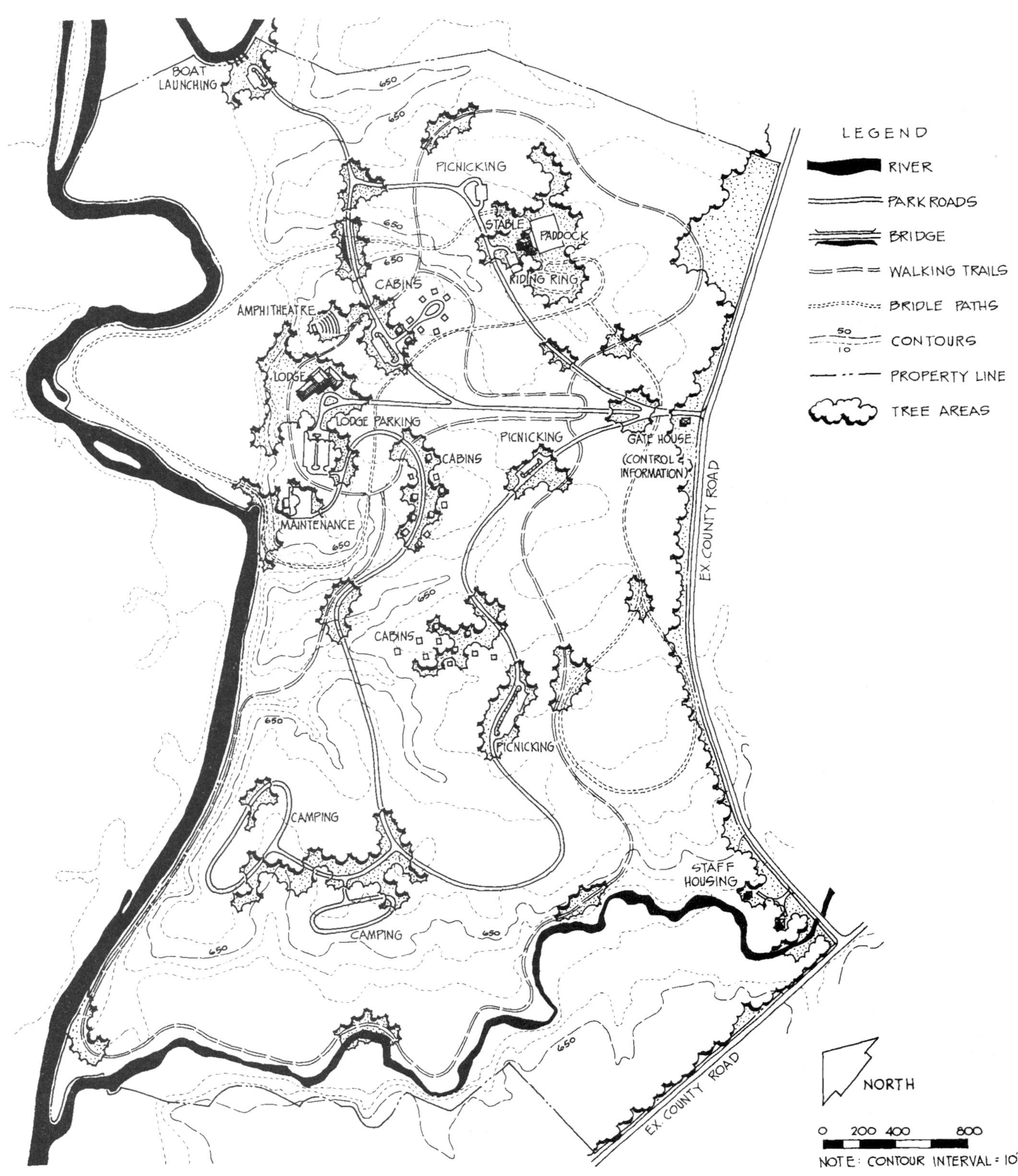

Figure 7-12
A state park master plan—solution 1.

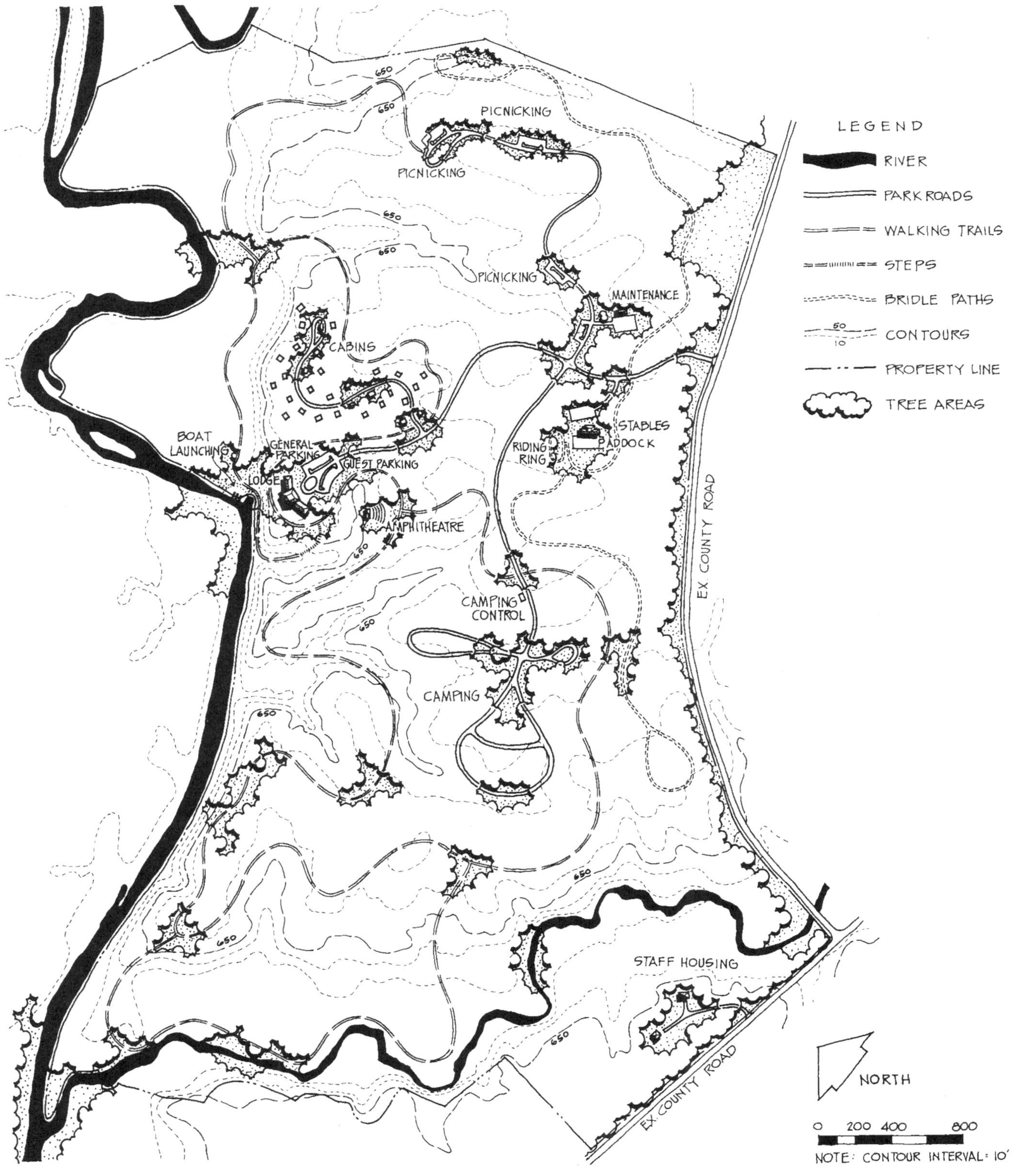

Figure 7-13
A state park master plan—solution 2.

Supporting your evaluation of these proposals, bear in mind the following information:

Total length of road
Solution 1: 16,100 feet Solution 2: 9900 feet

Total length of hiking trails
Solution 1: 18,200 feet Solution 2: 18,600 feet

Total length of bridle paths
Solution 1: 11,800 feet Solution 2: 6,000 feet

In evaluation, consider how these lines of circulation relate to land forms. Also consider relationships between different modes of travel—are they distinctly separate, intersecting, in conflict? Recognize that at this stage and this scale in a project, the objective is to get the big issues resolved and relationships correct. Don't get lost in details. In contrast to the data given above (which you should worry about now), elements too small to measure at this plan's scale would include items such as parking spaces, campsites, picnic tables (which you *shouldn't* worry about now). These will be verified later. Are they in the right places now?

Cases 5 and 6: Alternate Design Proposals for an Urban Parklet

Cases 5 and 6 again present alternate design solutions for a new park: this time an urban parklet to be developed on a corner lot, which is now essentially waste space. Once again you are provided with a program and initial criteria (Figure 7-14) and a site analysis (Figure 7-15). Go through the evaluation procedure for each of the design proposals (Figure 7-16 on p. 140 and Figure 7-19 on p. 141) and compare your conclusions. The plan scale is shown graphically. Use the perspective views of each site to help you mentally walk through the site, sense the space, and consider detail elements related to the total success of each design.

Supporting your evaluation of these proposals, bear in mind that virtually all data about the site can be determined at this scale. The actual number of car spaces, individual benches proposed, existing slope of the site compared to proposed slopes and grade changes represented by steps, types and areas of paving, and numbers and types of plants—all these elements are directly observable and measurable so that they may be used quite specifically in the plan evaluation.

From an overall observation:

Are issues and interests suggested by the site analysis ignored or violated?

From a more specific inspection:

Circulation—How are issues of circulation handled? Are there obvious routes offered or obstructions forced?

Views—What does the pedestrian and sitter see and are these views positive or negative?

Function—Seating: is the arrangement comfortable (enough space, shade), conducive to conversation; correct location? *Planting*: shade, barrier, beauty, correct scale and form, supportive of other functions? *Paved surfaces*: kind and quality for purpose, form supportive of function? *Furniture*: appropriate elements, correct locations, supportive of other functions?

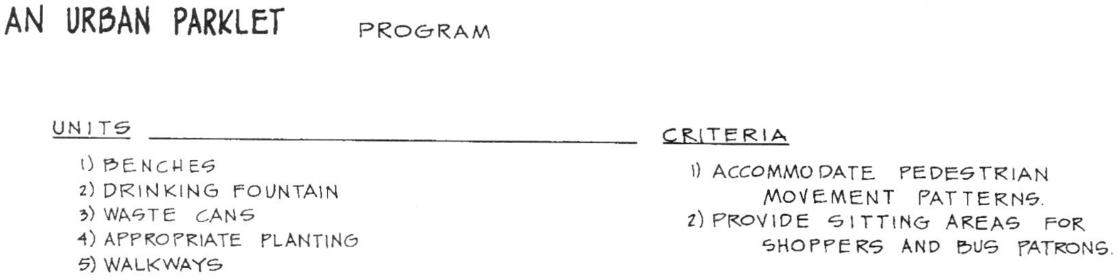

AN URBAN PARKLET PROGRAM

UNITS

1) BENCHES
2) DRINKING FOUNTAIN
3) WASTE CANS
4) APPROPRIATE PLANTING
5) WALKWAYS

CRITERIA

1) ACCOMMODATE PEDESTRIAN MOVEMENT PATTERNS.
2) PROVIDE SITTING AREAS FOR SHOPPERS AND BUS PATRONS.

Figure 7-14
An urban parklet program.

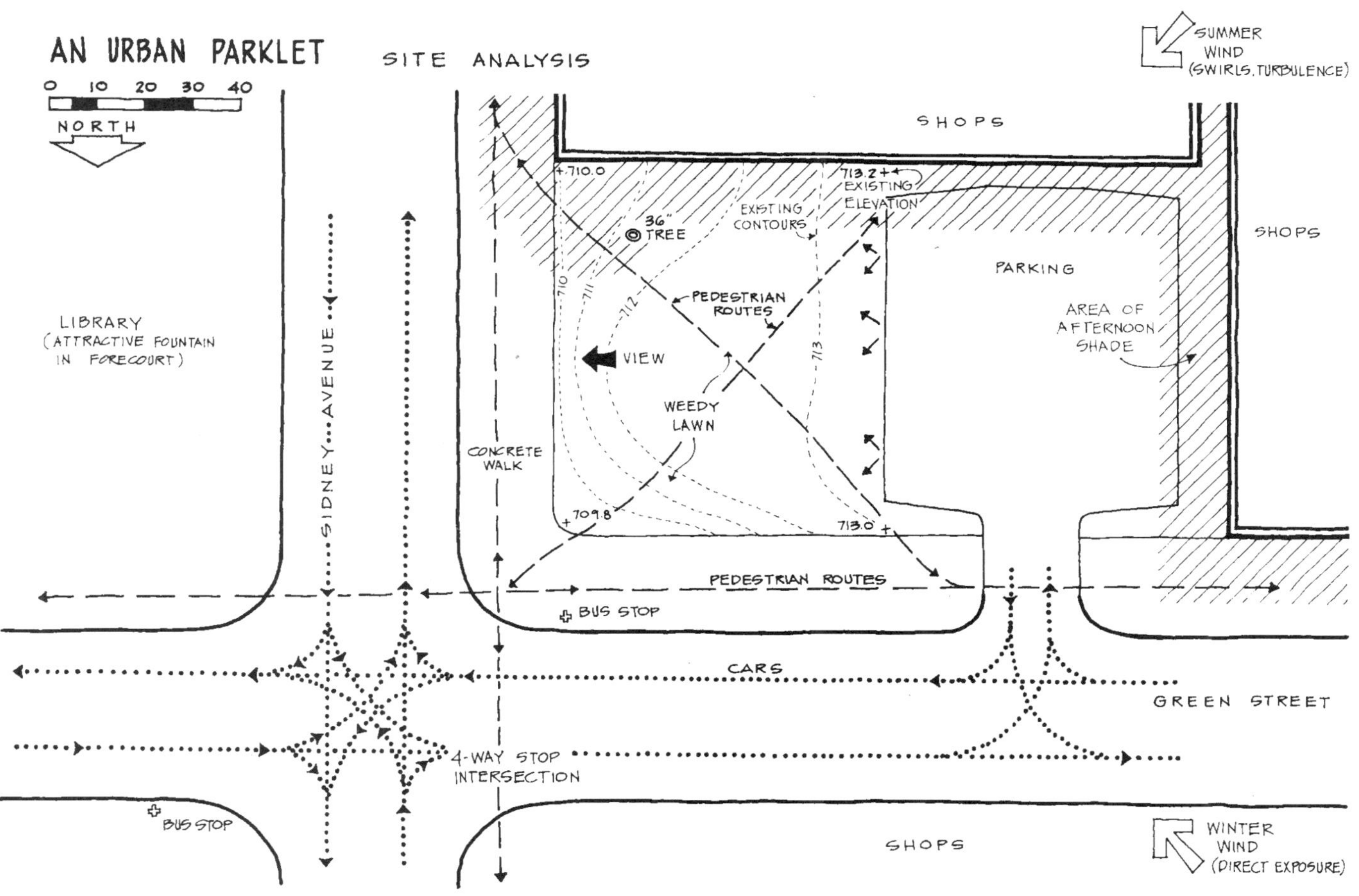

Figure 7-15
An urban parklet site analysis.

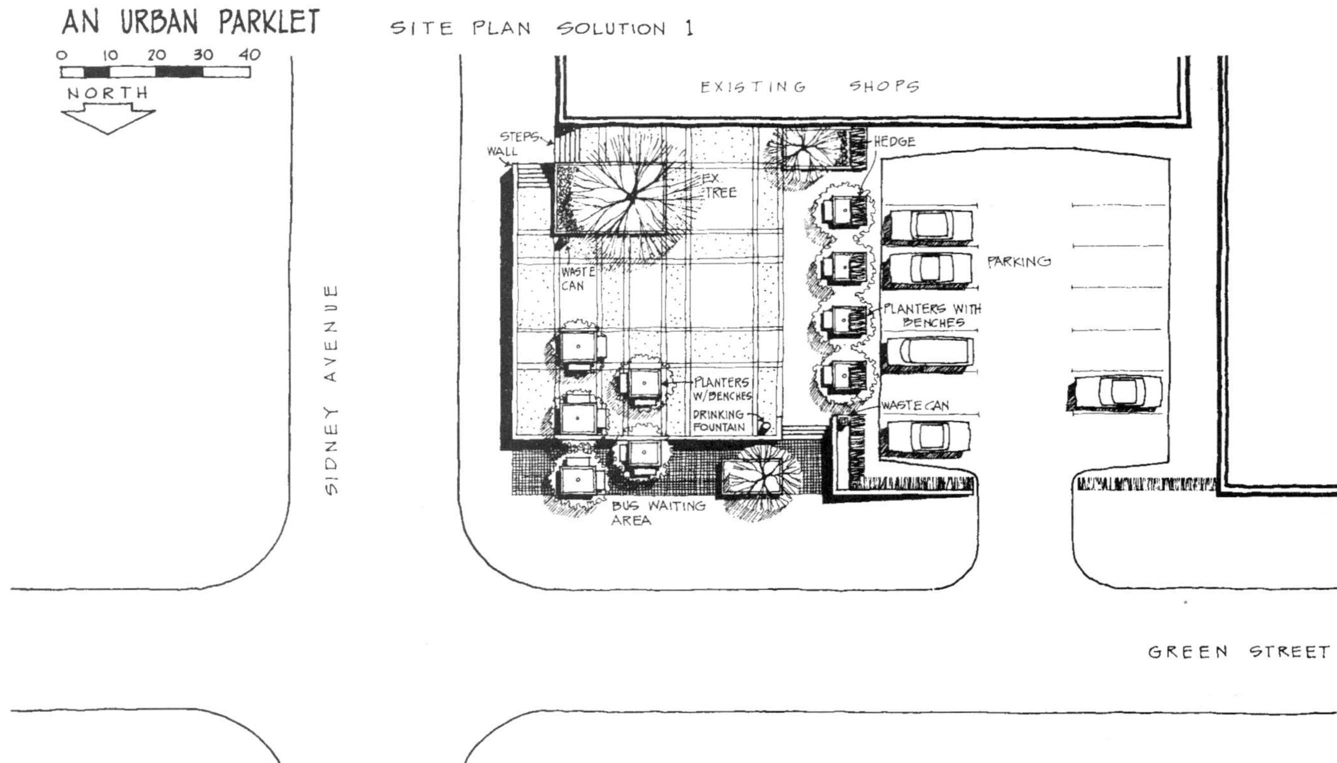

Figure 7-16
An urban parklet site plan—solution 1.

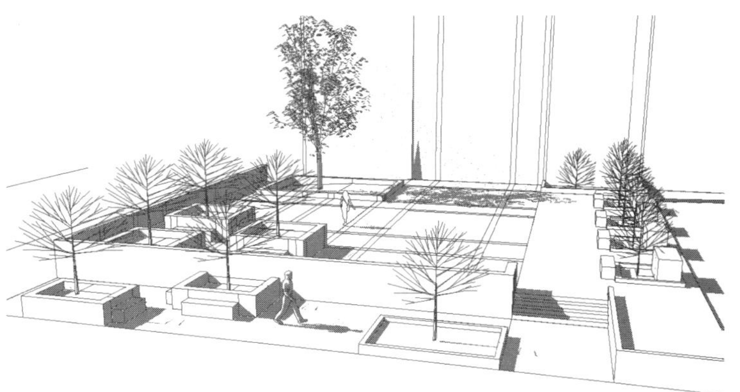

Figure 7-17
View of site from NW toward SE

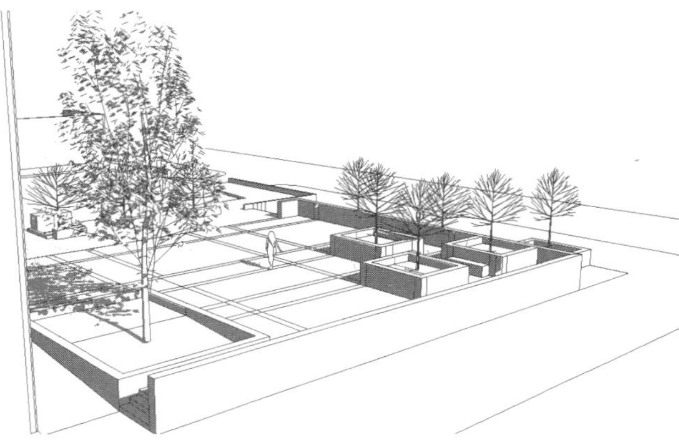

Figure 7-18
View of site from SE toward NW

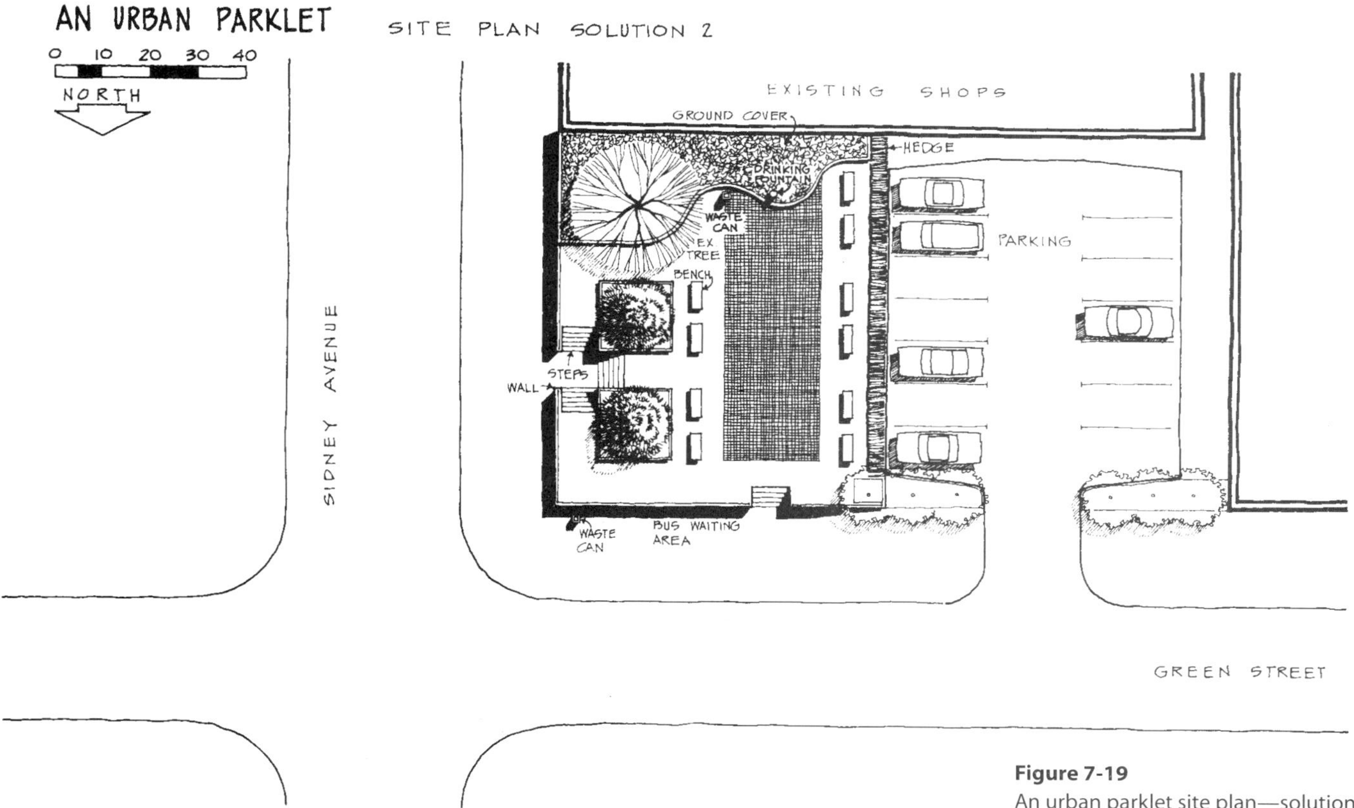

Figure 7-19
An urban parklet site plan—solution 2.

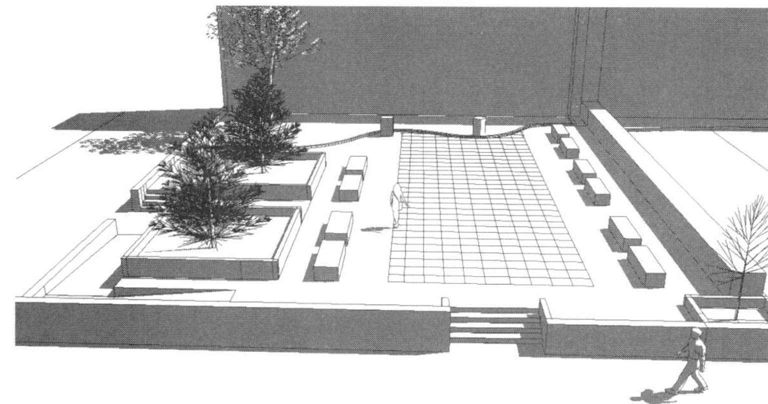

Figure 7-20
View of site from north toward south

Figure 7-21
View of site from east toward west

Aesthetics—Space well defined (sense of enclosure) and scaled to pedestrian activity? Elements of plan show unity and relationship to others, overall and in detail?

Park Redevelopment: Life Cost and Obsolescence

Two old-park case studies (Cases 7 and 8) will be presented to you next. Again, the analysis of the design is your responsibility. However, more background on the park, its location, and the people it serves is introduced than in the new-park studies, to help you evaluate the status of the existing design. The study of an existing park is necessarily a four-dimensional effort because the clock is already running. Since a park is already in place, the necessity of increasing the productivity of its resources motivates many of the procedures in each study. The study must consider the existing park for preservation or change as well as the creation of something new.

Life cost, as introduced in chapter 2, is determined by assessing the initial installed value of an element of a park plus the value of effort necessary to maintain this element throughout its theoretical useful life. In this application *maintenance* is defined as the work of keeping an element in good enough condition to serve its intended purpose. The greater the effort required to perform this service, the higher the ultimate life cost of the element. If the element is useful, the investment is a good one. If the element is not useful, the effort and investment are wasted.

Usefulness is the test of *obsolescence*. The test of usefulness must be applied to all elements of a park; this includes the plan itself. A plan can be obsolete even before construction if the functions it organizes will be obsolete when the work is done.

In order to evaluate the obsolescence of park plans, a "thermometer" test is offered (Figure 7-25 on p. 147), to be applied in Cases 7 and 8. This test is not a design evaluation. It is a subjective review of information about each plan, limited by the data available. These plans would probably be presented at 30 or 40 scale, and consequently the amount of detail that can be determined is limited. From this review, however, conclusions can be drawn as to whether the plan is obsolete, and the review deserves to be pursued further.

It is important to note that data are essential to the proper evaluation of obsolescence (and this is a further strong argument for constant devotion by a park agency to developing and using good record-keeping systems for all operations). The thermometer test is only a comparison and offers a basis for further investigation where questions remain. For other sites, more or different factors could be included.

Case 7: Redevelopment of an Existing Urban Park

In this case, modeled after a real and very similar situation, you will evaluate a redevelopment plan for an urban park. The park has an existing plan created for a user family of nearly a century ago. The genteel neighborhood of large and graceful homes occupied by single wealthy families was appropriately served by a large and very low-intensity park to be looked at, strolled through, discussed affectionately, and never touched. The existing plan now has to be reevaluated for a user population occupying those stately homes as apartment tenants at three or four times the original density. The need for recreation facilities has increased in the same proportion.

First, look at the background information that accompanies the existing plan (Figure 7-23 on p. 144). Based on this information, an obsolescence evaluation has been made (see the thermometer chart in Figure 7-25 on p. 147). We will now discuss the obsolescence evaluation point by point.

Congruence of needs and facilities involves a judgment on how closely the plans dovetail with the needs of the current users (Figure 7-22). These needs are learned by observing activities in the park, by interviewing users, or by surveying potential users. The park's facilities can then be evaluated in terms of the users' needs. Are the users there in spite of the park or because it "fits" them? The volleyball players are an example of people adapting facility to need.

The quality of the decisions made in this category is proportional to the range of users contacted. The need for data from all ages, social groups, and interest areas cannot be overemphasized.

The existing park is rated very low on congruence because it offers very little that the user doesn't have to adapt to use. The proposal (Figure 7-24 on p. 144) is rated high because it uses the park's framework to advantage in presenting settings for most of the needs identified.

To evaluate *supportive maintenance*, it is necessary to inventory all the work done to maintain the park and identify the activities each operation supports. Supportive maintenance would be regular and reasonable work directed toward continuing a desired condition. In the existing park, the lawn mowing and fertilizing in the area of the intensive volleyball wear is work done to support an obsolete activity (observation of the green as part of a park "picture") and opposes a use that fulfills a very real need (they even supply their own equipment!).

The existing plan is rated moderately low since it does foster the open-space activity occurring east of the pool. The proposed plan shows a better correlation of activities to physical facilities. In addition, required maintenance for the new facilities might be less than

AN EXISTING URBAN PARK PROGRAM

UNITS (OBSERVED*, REQUESTED•)

1) VOLLEYBALL *
2) PICNICKING / LUNCH *
3) OPEN SHELTER •
4) TENNIS •
5) MULTI-PURPOSE COURT/EXHIBIT SPACE
6) ENTERTAINMENT / SPECIAL PERFORMANCE AREA *
7) SEATING *
8) BUS WAITING *
9) BASKETBALL •
10) FIELD GAMES *
11) PLANTING / COLOR AREAS

CRITERIA

1) RETAIN USABLE SITE FACILITIES.
2) REDUCE COUNTERPRODUCTIVE MAINTENANCE.
3) ADAPT TO CURRENT USERS.
4) PROVIDE FLEXIBILITY FOR FUTURE.
5) PROVIDE INTERACTION OPPORTUNITY FOR USERS.
6) PROVIDE PEOPLE-WATCHING ENVIRONMENT.

Figure 7-22
An existing urban park program.

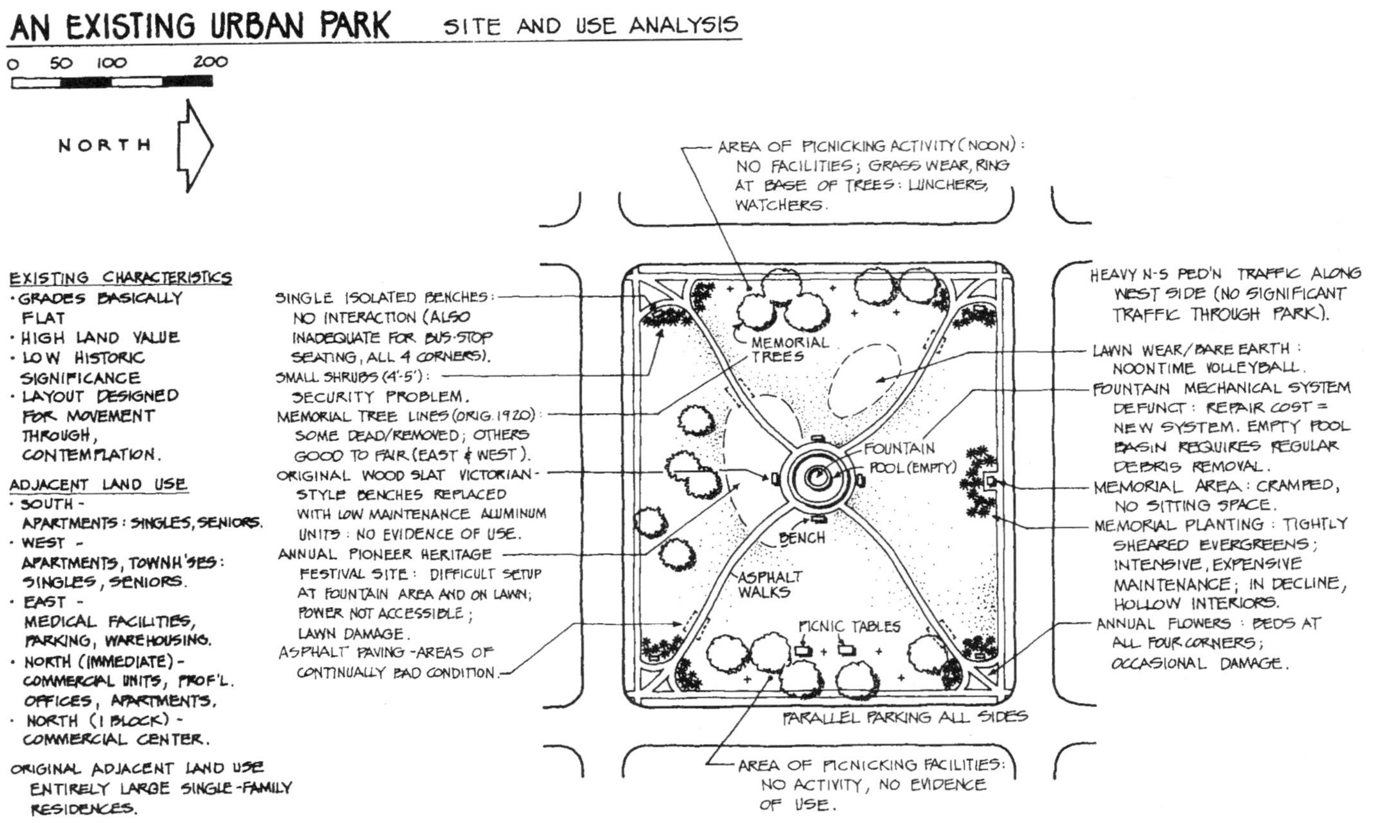

AN EXISTING URBAN PARK SITE AND USE ANALYSIS
0 50 100 200
NORTH
EXISTING CHARACTERISTICS
• GRADES BASICALLY FLAT
• HIGH LAND VALUE
• LOW HISTORIC SIGNIFICANCE
• LAYOUT DESIGNED FOR MOVEMENT THROUGH, CONTEMPLATION.
ADJACENT LAND USE
• SOUTH - APARTMENTS: SINGLES, SENIORS.
• WEST - APARTMENTS, TOWNH'SES: SINGLES, SENIORS.
• EAST - MEDICAL FACILITIES, PARKING, WAREHOUSING.
• NORTH (IMMEDIATE) - COMMERCIAL UNITS, PROF'L. OFFICES, APARTMENTS.
• NORTH (1 BLOCK) - COMMERCIAL CENTER.
ORIGINAL ADJACENT LAND USE ENTIRELY LARGE SINGLE-FAMILY RESIDENCES.
SINGLE ISOLATED BENCHES: NO INTERACTION (ALSO INADEQUATE FOR BUS-STOP SEATING, ALL 4 CORNERS).
SMALL SHRUBS (4'-5'): SECURITY PROBLEM.
MEMORIAL TREE LINES (ORIG. 1920): SOME DEAD/REMOVED; OTHERS GOOD TO FAIR (EAST & WEST).
ORIGINAL WOOD SLAT VICTORIAN-STYLE BENCHES REPLACED WITH LOW MAINTENANCE ALUMINUM UNITS: NO EVIDENCE OF USE.
ANNUAL PIONEER HERITAGE FESTIVAL SITE: DIFFICULT SETUP AT FOUNTAIN AREA AND ON LAWN; POWER NOT ACCESSIBLE; LAWN DAMAGE.
ASPHALT PAVING - AREAS OF CONTINUALLY BAD CONDITION.
AREA OF PICNICKING ACTIVITY (NOON): NO FACILITIES; GRASS WEAR, RING AT BASE OF TREES: LUNCHERS, WATCHERS.
MEMORIAL TREES
FOUNTAIN
POOL (EMPTY)
BENCH
ASPHALT WALKS
PICNIC TABLES
PARALLEL PARKING ALL SIDES
AREA OF PICNICKING FACILITIES: NO ACTIVITY, NO EVIDENCE OF USE.
HEAVY N-S PED'N TRAFFIC ALONG WEST SIDE (NO SIGNIFICANT TRAFFIC THROUGH PARK).
LAWN WEAR/BARE EARTH: NOONTIME VOLLEYBALL.
FOUNTAIN MECHANICAL SYSTEM DEFUNCT: REPAIR COST = NEW SYSTEM. EMPTY POOL BASIN REQUIRES REGULAR DEBRIS REMOVAL.
MEMORIAL AREA: CRAMPED, NO SITTING SPACE.
MEMORIAL PLANTING: TIGHTLY SHEARED EVERGREENS; INTENSIVE, EXPENSIVE MAINTENANCE; IN DECLINE, HOLLOW INTERIORS.
ANNUAL FLOWERS: BEDS AT ALL FOUR CORNERS; OCCASIONAL DAMAGE.

Figure 7-23
An existing urban park site and use analysis.

AN EXISTING URBAN PARK PROPOSED REDEVELOPMENT
0 50 100 200
NORTH
MINI-PLAZA
SHELTER
NEW TREE
EX'G TREE
GRASS/PAVER COMBINATION FOR TEMPORARY SHOW PLATFORMS, FESTIVAL SETUP
TENNIS WITH BASKETBALL
FIELD GAMES
PICNIC TABLES, REMOVED
CORNER PLAZAS: SEATING (CONC.) PLUS SEAT-PLANTERS FOR BUS-WAITING, PEOPLE-WATCHING.
VOLLEYBALL/MULTIPURPOSE COURT
EXPANDED MEMORIAL AREA WITH SEATING, CANOPY TREE PLANTING (SHRUBS REMOVED).
FLOWERING TREES

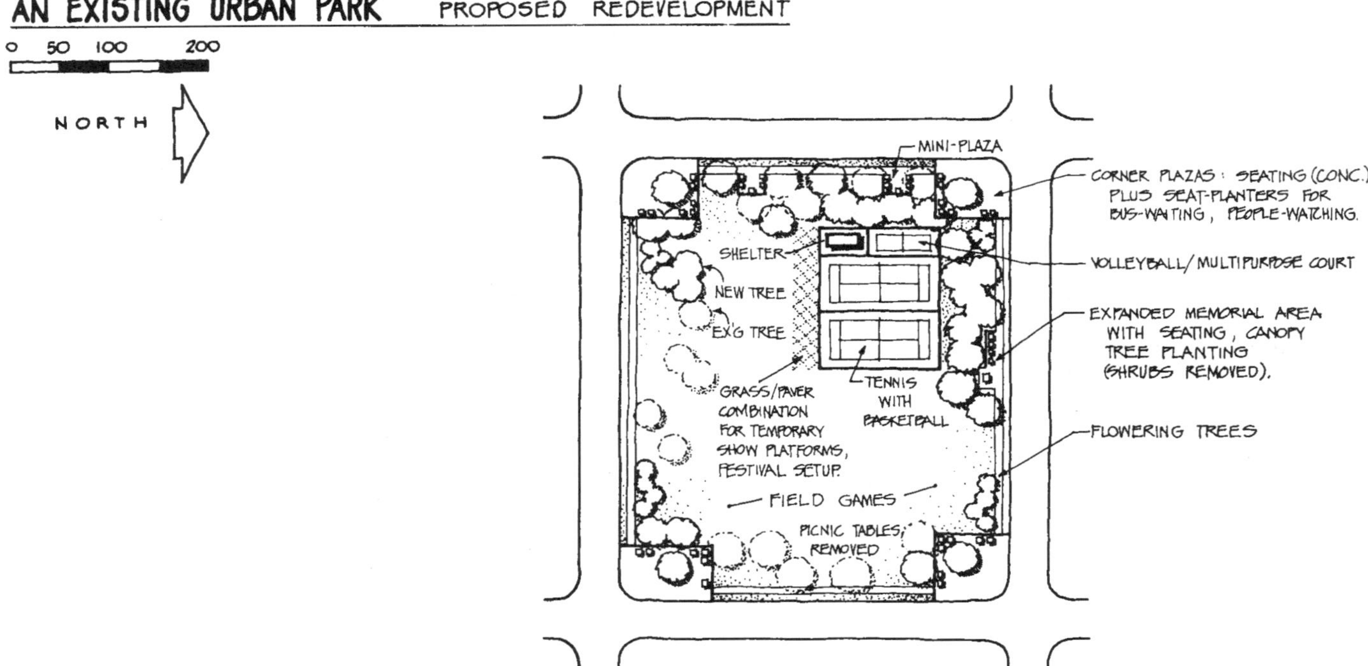

Figure 7-24
An existing urban park's proposed redevelopment.

for the present ones, for several reasons: less detail effort (elimination of shrub shearing); better material selections (the flowering trees for maximum results with minimum care, the grass pavers for the festival area, reducing the chance of damage and the resulting need for repair); and elimination of recurrent problems (removal of the asphalt paving, the pool with fountain, and the security-problem shrubs at the corners). The new plan also apparently makes feasible the use of larger equipment and simpler maintenance procedures.

Productive maintenance relates to activities plus elements. Productive maintenance would be specially scheduled work, a life-extending or renovating procedure—another installment payment on a good investment. It is not productive if the work is inefficient due to the design or materials used or if the activities it supports are obsolete. Restaining the shelter in the new plan for the first time after many years of good service would be productive maintenance. Shearing the memorial evergreens, for the second time in 6 months, to keep the ball shapes symmetrical is not productive maintenance. Amending the soil annually to provide a special acid condition for the flowering plants in the new proposal (assuming other equally effective materials are available) would not be productive maintenance.

The existing plan is a disaster in this category. The most painful example is the pool since, even though it has no function, it must still be maintained until a decision to eliminate it is made. The proposed plan eliminates the elements that don't offer program support and thus puts maintenance on a positive track. (Primarily, maintenance will consist of mowing with possible annual special service for paved areas, shelter, and some other portions of the site.)

Life potential for plants or construction materials is the average time the material can be expected to survive in reasonably good condition with reasonable maintenance. Life potential of plants or construction materials must be evaluated first in terms of the element's usefulness. If it is obsolete for its original use and there are no other arguments for keeping it, the life potential is zero.

If usefulness is established, can reasonable maintenance work maintain the element's function and appearance? Maintenance requirements for the fountain in the center of the existing park were no longer "reasonable" when the plumbing system aged to the point that overhauls of the mechanism became necessary twice a year, imported parts were needed, and plumber's rates exceeded those of the agency's director.

If reasonable maintenance work seems sufficient for the foreseeable future, then is the cost of that maintenance for the element's anticipated life less or greater than the cost of a new equivalent? At any point a new acquisition offers the elimination of all the old problems and the advantage of some new opportunities. The memorial evergreens in the existing park, assuming they could be tolerated with only one shearing per year, would still in a very short time consume a maintenance allowance that would permit buying new plants selected to grow within the design limits without requiring special pruning.

The existing plan is rated low on both its plantings and its construction elements. The plants are all in stages of advanced age. The high-maintenance materials have the lowest functional value. The medium-height shrubs creating the security concern at the corners are doubly bad, since any work on them not only costs money but further exacerbates the problem. The proposed plan scores well in this category because its plant materials are simple, big, and useful. The design groups them in most places, thus providing backup so that the individuals are neither so important nor so vulnerable. The construction materials are basically all simple, with none of the complexity of the pool or the vulnerability of the narrow asphalt paths.

Future flexibility is not easily evaluated. What will be needed in the distant or near future is impossible to determine. It can be assumed, however, that a facility with a single, very specific function will not be very useful if that function becomes obsolete. A facility like the pool with central fountain offers no possibilities for change. However, a paved area painted with current game patterns and identified as a multipurpose court can later be repainted and continue its useful life in a new capacity. Likewise, an area designated as playfields can, if "insurance" space has been included in the plan, expand or contract as necessary to accommodate new activities.

Other than the open lawn areas, the existing plan offers little flexibility and is scored low. Its facilities are restrictively designed to do only one thing and are not amenable to change. The proposed plan scores well because it provides facilities that could be easily adapted to new uses without much more effort than a title revision. The design also makes prudent use of the park's space by not committing the center of the site to the paved court complex. The remaining open lawn area permits multiple activities with some buffer between.

Sentimental and historical value are confused with each other in some cases, especially by the user who remembers "that day in the park" with teared eye and lumped throat. Sentimental attachment to a park is expressed in user reaction against *any* change. The more important consideration is that of historic elements, either natural (as an unusual geologic formation, plant community, or event—like Capistrano's swallows) or sociopolitical. If significant enough, these elements don't become obsolete simply because time is frozen for them.

There is evidence of sentimental concern for the existing park on the part of some users (and consequently a high rating). There is no evidence of any historic significance except for the memorial structure and the trees. Both of these elements are maintained (in fact enhanced) in the proposed scheme. The historic rating is consequently equal for both plans. The attraction of a new development could be weighed against the sentimental attachment to the existing one through a further effort at user contact in the surrounding area.

Summary

The judgments recorded in the thermometer chart in Figure 7-25 are subjective assessments of the limited data available. The proposal scores well for many reasons, most of which are summarized as increased productivity from the resource. The existing facility would simply be used more effectively for better purposes.

Rather than a high-low range, it would be more descriptive in this sort of evaluation to provide a numerical score (ranging from, say, 5 for a condition that meets the criteria to 1 for an unsuitable or obsolete condition). A low total score would indicate that a plan is obsolete; this method also permits a quantitative comparison of the obsolescence of rival plans. As more data are assembled about a park, the quality of evaluation improves; and the evaluation can be more confidently presented to the clientele knowing the direction suggested can be supported by the facts. For the next case study, you will be required to do your own thermometer evaluation, this time of an existing plan and two alternate redevelopment plans.

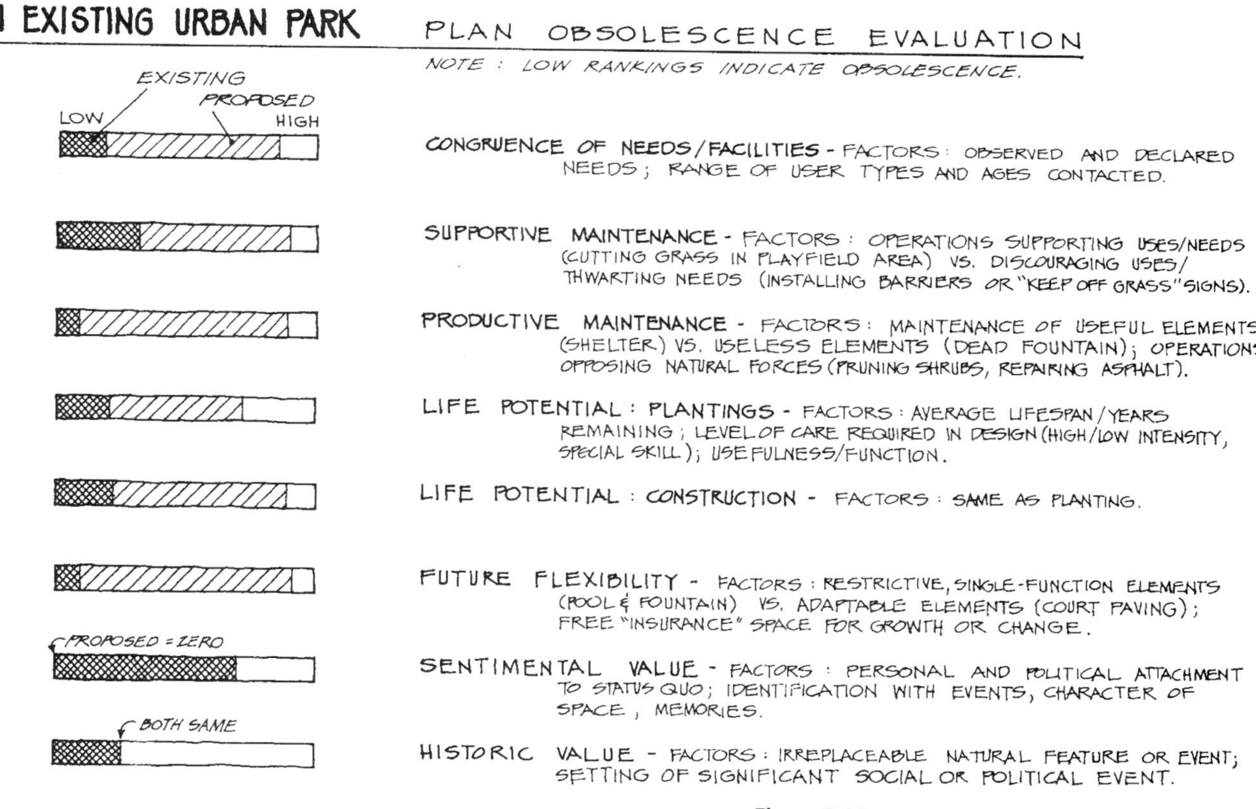

Figure 7-25
An existing urban park's plan obsolescence evaluation.

Case 8: Redevelopment of an Existing Regional Park

This study is also based on a real site with considerable significance to its agency because of its central location in the urban area and its large size. The program (Figure 7-26 on p. 148) is a distillation of data obtained through broad public contact. The incorporation of commercial elements like the waterslide has been reviewed at great length and accepted as a positive direction for the agency without legal or principle conflicts.

Initially (about 1900), the population of the surrounding neighborhood was low to middle income and of European heritage. Through the 1930s popular activities were car touring, picnicking, and some field sports. Other uses grew by vogue or accumulation of tradition. The park continues to attempt to serve nearly everyone for everything.

The current population in the local area is still predominantly single-family. However, the park now serves a much wider region because of the ease of access to it. Major roads on the edges of the park connect it to the entire urban area. Consequently a broad spectrum of users considers this their park.

In addition to those with a strong desire for the commercial activities programmed, there is a very lively population interested in the existing natural resources of the site. All-season hikers, bird watchers, and naturalists enjoy the park. There is a strong garden club in the community with the financial and organizational ability to actively

AN EXISTING REGIONAL PARK PROGRAM

UNITS (SOURCE: SURVEY OR OBSERVATION)

ENTERTAINMENT FACILITIES
1) WATERSLIDE
2) OBSERVATION TOWER OR EQUAL LANDMARK
3) NEW CARNIVAL RIDES
4) CHILDREN'S THEATER
5) SEPARATE, EXPANDED CHILDREN'S PLAY AREA
6) RETAIN / IMPROVE:
A) ZOO
B) BAND SHELL
C) LAGOON
D) ROSE GARDEN
E) GAZEBO
7) ZOO SERVICE BUILDING
A) EXPAND FOR MORE MAINTENANCE/ STORAGE SPACE (NOW IN CONJUNCTION WITH HEADQUARTERS SERVICES).
B) SEPARATE H.Q. (TO ANOTHER SITE); EXPAND FOR ZOO SUPPORT AS INTERPRETIVE/EDUC. CENTER.

EDUCATION FACILITIES
8) CONSERVATORY ADDITION FOR ADULT AND SPECIAL EDUCATION CLASSES.
9) OUTDOOR TEACHING FACILITY IN CONJUNCTION WITH THE CONSERVATORY.
10) IMPROVE HISTORIC CABIN SITE FOR PROGRAMS.
11) GARDEN FOR MAJOR EXHIBITS, FESTIVALS; SUNKEN FORM REQUESTED FOR DEFINITION OF SPACE AND CONTROL.

SUPPORT FACILITIES
12) EXPAND PARKING 20%-30%
13) CONCESSION STRUCTURE(S)

CRITERIA
1) UTILIZE EXISTING FACILITIES TO SATISFY NEW NEEDS.
2) PROVIDE MAXIMUM SEPARATION OF CONFLICTING UNITS (NOISY/QUIET, EDUCATION/ENTERTAINMENT, ETC.).
3) PROVIDE MAXIMUM EASE OF CIRCULATION
4) PROVIDE EFFICIENT CONTROL FOR COMMERCIAL ELEMENTS (WATER SLIDE, FEE PARKING, ETC.).
5) IMPROVE POTENTIAL FOR EFFICIENCY OF SERVICE, MAINTENANCE.
6) RETAIN MAXIMUM VALUE OF NATURAL RESOURCES ON SITE (WOODS, WATER, ANIMAL HABITAT).
7) REDUCE COUNTERPRODUCTIVE MAINTENANCE.

OTHER FACTORS
1) USE INTENSITY IS LOW ON TENNIS COURTS. SINCE MAINTENANCE CONTINUES AT NECESSARY LEVEL, OPTIONS TO INCREASE USE OF SURFACE SHOULD BE EXPLORED.
2) HEADQUARTERS COULD BE RELOCATED OUT OF PARK IF REDUCED CONFLICT AND INCREASED USE POTENTIAL WOULD RESULT.

Figure 7-26
An existing regional park program.

support development efforts. Equally active are several groups concerned about historic resources and preservation in the community. Community entertainment is supported by symphony and theater support organizations, each with interest in the park as a location for major public performances.

Other community resources in the area include schools from elementary to high school. The school populations use the park on a scheduled basis for activities such as cross-country meets or performances. The baseball fields have been reduced to inactive status by construction of new fields with bleachers and full services at the high school campus. Other student activities tend to be somewhat more casual. A significant population of teenage cruisers enjoys the park in the evenings (at least parts of it).

In evaluating the two redevelopment proposals (Figure 7-29 on p. 151 and Figure 7-30 on p. 152), it is important to recognize some major guidelines to which the proposals responded. Alternate B (Figure 7-30) responds to the strong concern expressed by a faction of the board about the safety of the park's road system. The existing road system is recognized to have several hazardous areas (Figure 7-27), the result of attempts to accommodate shifts in circulation patterns as

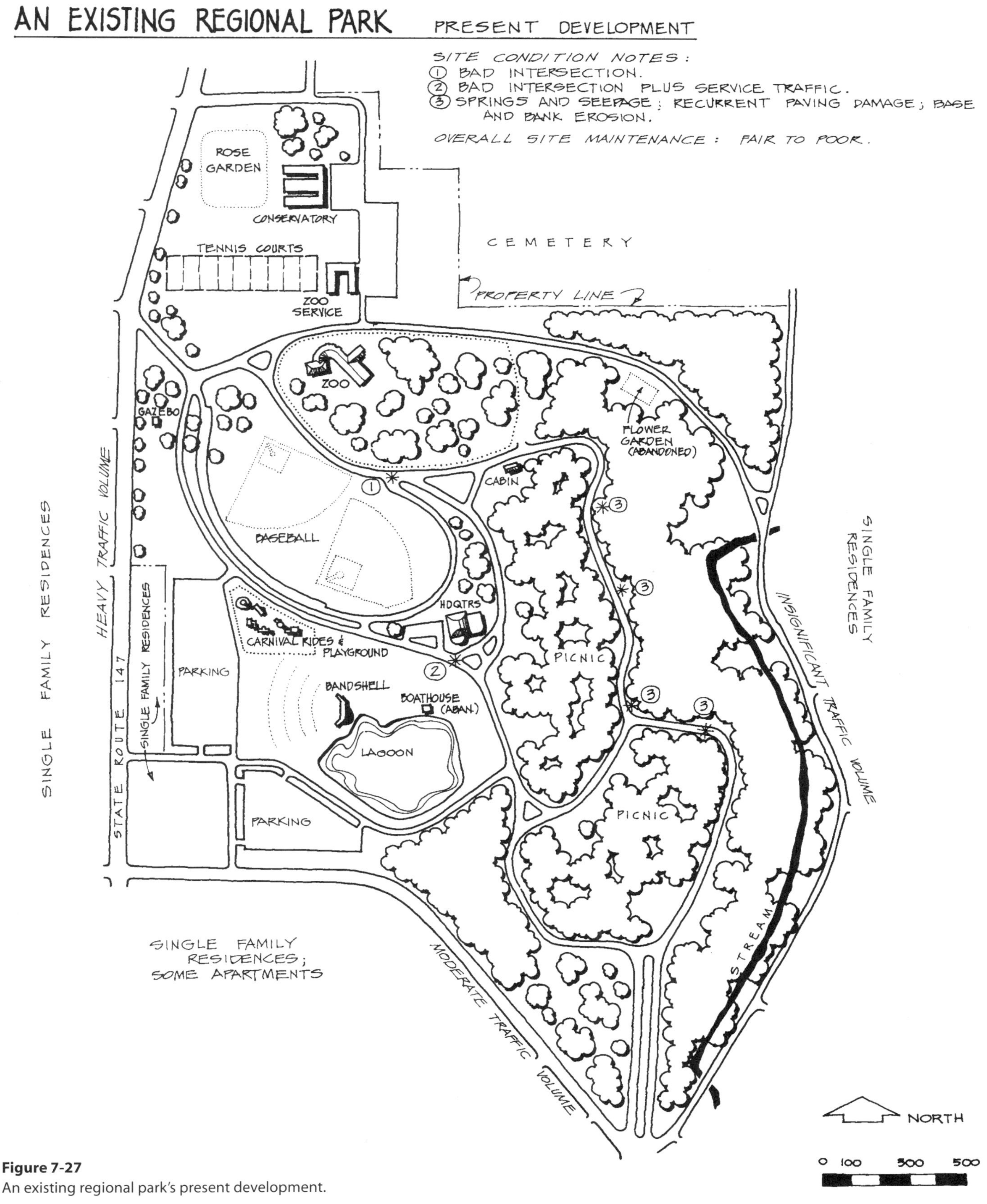

Figure 7-27
An existing regional park's present development.

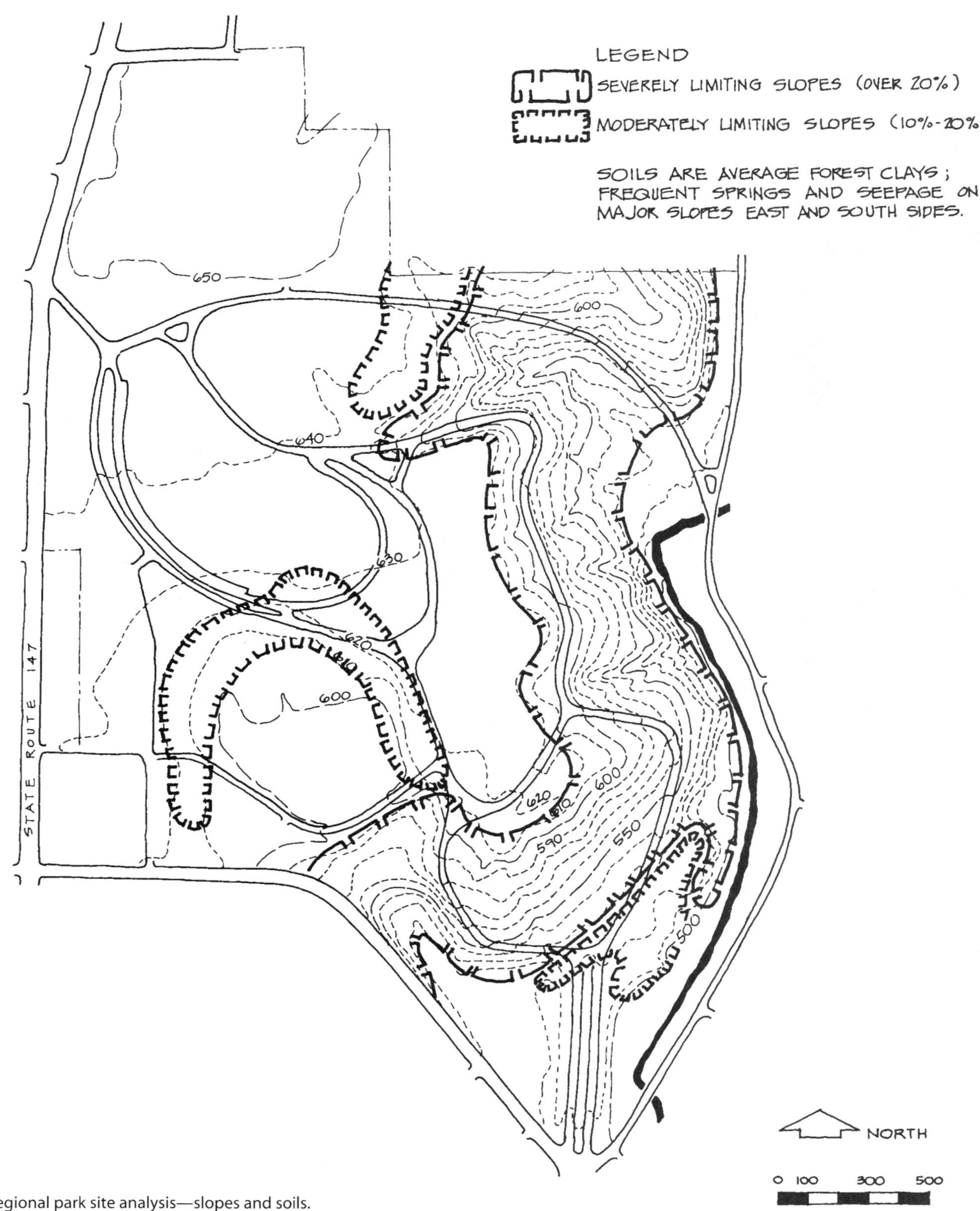

Figure 7-28
An existing regional park site analysis—slopes and soils.

Figure 7-29
An existing regional park's proposed redevelopment—Alternate A.

Figure 7-30
An existing regional park's proposed redevelopment—Alternate B.

new uses grew on the site. Notice, for example, dangerously angled intersections, as well as head-in parking along roads necessitating departing cars backing into traffic. The developers of both schemes were aware of the importance of supporting the commercial elements programmed, and consequently both attempt to provide the best possible solution to vehicular circulation, parking, and people movement.

Consider the setting and background, then develop your own thermometer or other device to evaluate the obsolescence of the existing plan and the potential of the new plans. Then evaluate the design proposals in detail. Bear in mind that the plans offered do not give by any means the only answers possible. Try to do more than select a winner and a loser. Evaluate your own suggestions for revisions and additions.

To guide the evaluation of these two design proposals, put yourself in different roles, for example: as a visitor or user of the park, as the manager of park operations or director of maintenance, and as the design team leader looking at suggestions by your staff. From these points of view, you might consider the following issues:

Visitor/User

- *Circulation*: If you're driving will you be able to navigate well to various targets? Are the road layouts safe and simple, or are there conflict points or redundancies? If you're walking, are pathways convenient and are point-to-point routes safe or in conflict with other circulation (visitor cars, maintenance trucks, etc.)?
- *Driving or walking*: Do you feel the park is biased toward either of these?
- *Parking*: Is parking visible; convenient; safe to enter or exit?
- *Uses, activities, functions*: Are all the things you want in this park represented? Too much? Too little?

Manager/Maintenance Director

Have existing maintenance problems been recognized and resolved? Are service and storage areas consolidated or separated? Are uses requiring similar services consolidated or separated? Are roads adequate for circulation and service access without redundancies? What level of maintenance will be generated by the park's elements? Can the maintenance be mechanized and done in large simple steps or will the work have to be done by intensive hand labor? Will the maintenance have to be frequently repeated?

Design Team Leader

Are elements sensibly related:

- to each other—compatible uses adjacent, user ages together, noise zoned?
- to the site—in shade, in sun, slopes or flat, best orientation?

Is access reasonable, safe, a good experience? Are intended functions accomplished by design or placement of elements? Esthetically, are forms and spaces appropriate, pleasant?

Finally, as a "committee," use the obsolescence study technique to evaluate which of the designs hastens its own aging process by more or less success in the design decisions you have just evaluated.

Obsolescence Studies: A Caution

The evaluation of the obsolescence of any site—whether for a park or an industrial complex—is not a useful exercise if performed in isolation. The study is justifiable only to the extent that it is used as a means of comparing one property to another: *Which is more nearly obsolete?*

Successful development of obsolescence studies will be in direct proportion to the continuing effort by any agency to improve both the quantity and quality of the records it maintains about all its operations. Why bother? Because until such time as agencies find they have unlimited funds, there have to be ways to decide which projects deserve the available funds. This is at least a way to start.

Case 9: Development of a Linear Park—A Composite Trail Study

To introduce this study, following is a capsule of principles and conditions involved in the design and development of linear parks.

Some terms need to be introduced or defined in this application. *Linear parks* are narrow strips of recreational lands that often follow a linear feature such as creeks, rivers, railroad or utility rights-of-way, or highways. They also may be developed along utility rights-of-way or on leftover linear open spaces in the community. Linear parks typically include trails, trailheads, and open space. *Trailheads* are places for people to meet or for families to get organized before hitting the trail. Existing parks often become trailheads for new greenways. *Intersections* are points of contact of two or more trails or multiple modes of movement (bikeways, motorized vehicles). *Nodes* are points of special interest, significant viewpoints, or significant extra space.

Linear parks often serve as connectors between parks. Greenways in many instances may seem concurrent with linear parks, except that greenways need not include trails. Greenways that are lands set aside for watershed protection or wildlife habitat and movement corridors are *not* considered linear parks.

Trails are essential to a linear park, but trails do not make linear parks. Trails that circumnavigate a park do not constitute a linear park. Trails that parallel city streets are wide sidewalks, not linear parks. On-street bike lanes are neither trails nor linear parks. While linear parks may have segments that are only wide enough for trail connections (a twenty-foot right-of-way should be considered a minimum), the greenway qualities of a linear park are achieved when its width is in excess of fifty feet (and two hundred feet would not be too wide). Linear parks may be a simple corridor connecting two parks just a few blocks apart or a complex system of trails and greenways throughout or between communities.

Linear parks are different from areal or rectangular parks in that they may serve as corridors of movement throughout the community, offering transportation as well as recreation. The trails in a linear park may also be interpretive, providing education about the history of a former railroad or the importance of nature along a river corridor.

Planning for Linear Parks

Land that is greater in length than in width would presumably qualify as a linear park, but what makes a *good* linear park? A good first step is to develop a community or regional plan for an entire system of linear parks (see Figure 9-2). It is best to identify all possibilities for linear connections. General plans can develop later into specific plans. Citizens of the community need to be part of the planning process for a complete park system. They need to buy into the vision.

When the vision is defined, priorities must be itemized. Priorities depend on availability or cost of land, the ability to secure funds for the project, and the attractiveness of each linear park segment that is likely to be feasible, successful, and popular with the people of the community. A linear park connecting a school with a neighborhood park is a much better choice for development than an obscure area along a creek at the edge of town. The development of a linear park between two existing parks may not require trailheads.

Bigger is better. Generally, longer trails are better than shorter ones, given the ability to fund their construction and maintenance. Up to a point, wide corridors are better than narrow ones because they will accommodate more users, that is, more park users and shared use with wildlife.

Urban trails provide more immediate use to more people. Urban linear parks provide connections between parks, neighborhoods, workplaces, schools, and shopping areas. Rural linear parks or trails are more likely to connect adjacent communities, provide opportunities to get away, to enjoy open space, to view wildlife, and to travel farther.

Design for Linear Parks

Design for linear parks is not that different from that of other parks. Many of the elements are the same, but the emphasis is different. A program must be established, incorporating at least a list of intended users and uses. Design issues are concentrated on the trails. Urban trails are ten to fourteen feet wide, whereas rural trails may be eight to ten feet wide (all still needing at least the twenty-foot minimum right-of-way). Paved trails are more common in high-use urban areas; they accommodate walkers, strollers, wheelchairs, rollerbladers, and all users needing a smooth surface. Rural trails may be paved or unpaved. Crushed stone or other appropriate aggregate is probably the most common surface for rural trails. The stone or aggregate should be local, easily available, and compactible to provide a stable surface.

Rural linear parks may have pedestrian/bike trails as well as equestrian trails. Horses and hikers are usually incompatible trail users. Horses heavily impact trail surfaces and leave unwanted deposits. Horse riders often don't see (or smell) this incompatibility. If there is just one trail going to the bottom of the Grand Canyon, there is no other option, but in most places space is available for separate trails. Horses prefer unpaved trails. Wide, grass-covered trails are best. Horses are safer on separate trails, away from hikers who might spook them.

The design of a linear park is the creation of a sequential, linear experience. It is a perfect opportunity to present the landscape to the users as a sequence of educational and interpretive opportunities. The journey becomes the all-important experience of the linear park. The sequence includes areas of dark and light, sun and shade, enclosure and exposure, wind and calm, closed and open views, and a variety of visual pleasures. These design manipulations can make even a very straight rail trail seem interesting and enjoyable.

Facilities along or within a linear park are usually found at intersections, nodes, or trailheads. These are the places for parking, picnic facilities, drinking fountains, restrooms, and lighting. Such places are accessible to service and maintenance equipment and personnel.

Figure 7-31
Trail information sign.

These facilities should be identified by clear and consistent graphics (see Figure 7-31), which should include the advance identification of elements ahead and the distance to them.

Management of Linear Parks

Management activities include maintenance of a clean and safe park environment. Maintenance of a clean trail may require litter pickup by park staff or volunteers. Graffiti-free trails and clean restrooms are important to most users. Physical safety and personal security are important on the trails. (See Figure 7-32.) Physically safe trails are smooth, free of potholes, and clear of overhanging tree branches. Design features such as fences, bridges, park furniture, and signs must be durable to remain safe.

Safety in terms of personal security is an increasingly important aspect of park design and is an even greater concern in linear parks. The dangers associated with trails may not be any greater, but public perception is that linear parks are more susceptible to criminal acts. Unfortunately, many linear parks have a past life as an abandoned rail corridor or other hidden area where troublemakers found refuge.

Therefore, the key to making places safe *for* people is in populating the places *with* people. People using linear parks make the parks uninhabitable by potential criminals. Yes, crimes have occurred in broad daylight in Central Park with hundreds of people in sight, but this is the exception. The security of linear parks rests with the task of keeping them in usable, attractive condition and promoting their use. Some linear parks are monitored by a cadre of volunteers that ride or walk the trails to provide assistance to the tired or injured and provide ready communication to higher authorities if the need arises.

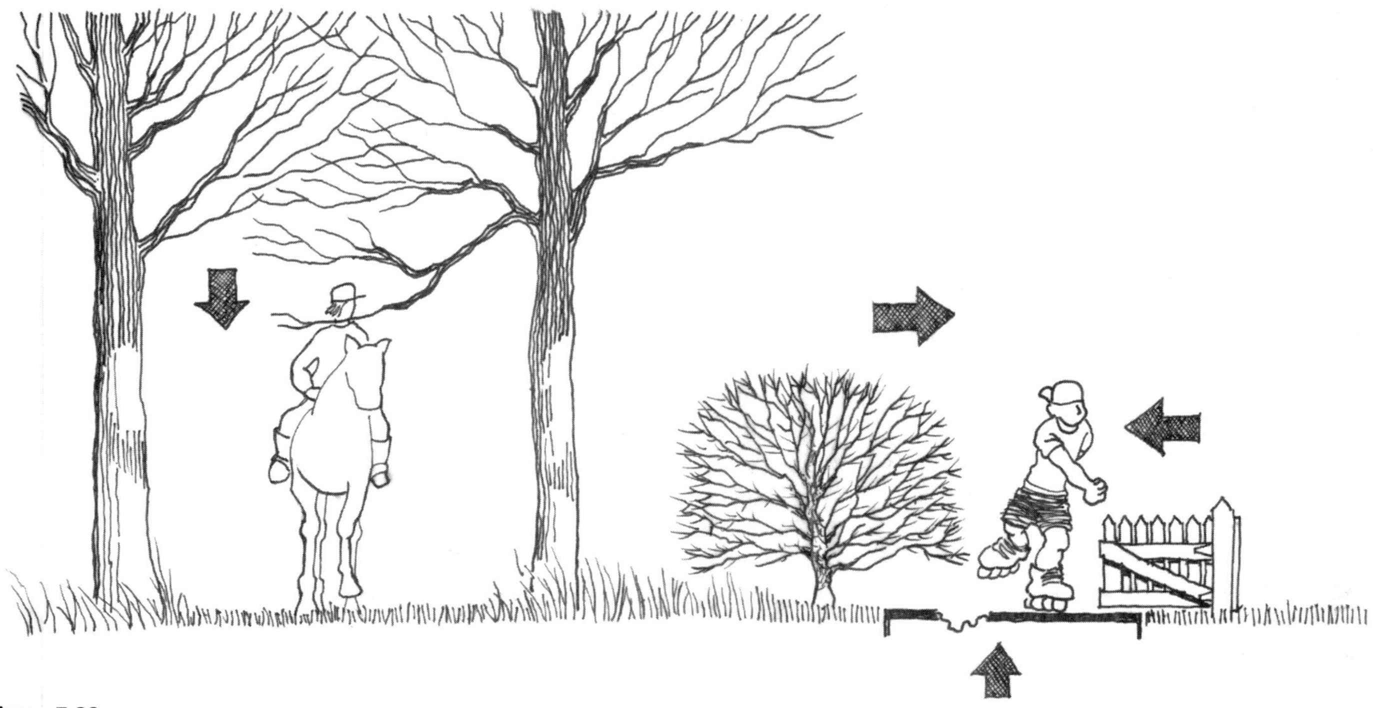

Figure 7-32
Trail hazards above, below, and beside.

Many linear parks have been developed in recent years to greatly expand the range of experience base from which all planners can draw. Whether park planner or park board member, you should find every opportunity to visit numerous linear parks to determine which characteristics, design techniques, and management strategies best suit the needs and desires of your own clients or communities.

In chapter 5 on Plan Interpretation, recall that the conclusion recommended that you try to "picture yourself walking inside the plan." Now for the next step beyond: Completing this specific case study assignment necessitates identifying your own linear park/trail, whether existing or planned, and visiting the full length of the site with the following checklist in hand. Just as some of the case studies require evaluation and reevaluation on paper, you'll have to make more than one trip along the actual route in this case. You'll need a paper copy of a map of the route. First, just travel the entire route after thoughtfully reviewing all the checklist items. Then repeat the journey, actively marking your map with item numbers from the list. For each item in the checklist, simply note the number on the map with a plus-sign if the element or characteristic exists or is positive (attractive, comfortable, etc.). Conversely, use a minus-sign by the item directly on the checklist itself if the element doesn't exist or if a characteristic is negative (ugly, dangerous, etc.). Consider the overall "personality" of the total site by the abundance or lack of marks for various items. Does the resulting description of the park actually fit the intended user group or fit into the overall intention of having this kind of park in the park system?

Consider whether there seems to be too much or too little of certain things. Are there opportunities for changes, additions, extensions? Can different users be encouraged through education programs? Can land be added or adjacent conditions be modified? The greatest opportunity in development of a linear park is the potential flexibility of route selection to contact or avoid specific elements. Consider using it to maximum advantage.

Linear Park—Checklist for Case Study

1. Program
   a. Users
   b. Uses
2. Esthetic experiences, contrasts
   a. Sun/shade
   b. Enclosure/exposure
   c. Dark/light
   d. Wind/calm
   e. Views—closed/open
   f. High/low
   g. Straight/curving
   h. Other ________________
3. Esthetic experiences, sequence
   a. Sun/shade
   b. Enclosure/exposure
   c. Dark/light
   d. Wind/calm

e. Views—Closed/open
f. High/low
g. Straight/curving
h. Other ________________

4. *Facilities* (list at each location—parking, picnic, water fountains, restrooms, lighting)
   a. Trailheads
   b. Intersections
   c. Nodes
5. *Safety—physical*
   a. Surface condition
   b. Side obstructions
   c. Overhead obstructions
6. *Safety—security*
   a. Frequency of use
   b. Active programs of use
   c. Contact with security personnel
   d. Visibility of adjacent areas
   e. Telephone accessibility

But What About All the Other Kinds of Parks?

Oops. We didn't do a case study of the kind of park you're going to be responsible for? What kind of park might you have to evaluate that isn't in the case studies?

- *Bigger:*
  Yellowstone II
  State of Rhode Island
- *Different Site/Setting:*
  Totally urban, office/building-bounded, all hard-surface, no planting
  Totally underwater, no hard surface, *really* no "planting"
- *Radical New Use:*
  Kangaroo park (nobody thought there would be dog parks either)
  Racing and special-event park for two-wheel gyro-stabilized scooters
  Extreme sport facilities
- *Radical New User (see Radical New Use above)*

In fact, your special park has already been covered. Consider what's common about every case study offered:

- *Program*—a use/purpose description had to be established.
- *Site*—which in other cases might have been multiple sites from which the best was to be chosen.
- *Inventory and analysis* of the site—resulting in the best places chosen for the program elements.
- *Relationship analysis*—to determine how program parts could best fit together.
- *Design concept and refined plan*—in more or less detail, depending on the *scale* of the project.

- *Long-term consequences of project development and use*—the incorporation of the project into the park system and the maintenance of it through its life.

Now consider what's different about the park types listed above (and not case-covered):

- *Scale*

That's it. No matter how big and complex the park may be, it has all the same elements to guide its evaluation. Eventually all the biggest issues of land use, element relationships, grading of mountains, and mowing of meadows, will be followed by issues of the specific plant in the bed or the fasteners of the bench legs.

Your evaluation process may require including a special advisor to tell you how high and how far a kangaroo can leap before you can be sure the space designed is adequate, but the same is true for the baseball field in the most traditional of public parks. If you never saw or played baseball, you'd need advice from an expert to know if the outfield fence line is too close. But all the procedures of the evaluation process would still be applied to both parks. The state park project, the largest of the case study sites, will be evaluated eventually at the level of detail appropriate to the placement and construction of the picnic tables, essentially the same issues to be considered for evaluating the tiny urban parklet. The essential things to remember are the sequence of the evaluation (first things first) and the importance of at least considering all the steps. The details don't matter if the park is in the wrong place.

8 Establishing the Client–Consultant Relationship

Being the Client

Unfortunately, a park doesn't just pop out of the ground in full glory. There has to be an agency responsible for the park. And the agency has to have somebody in charge—usually a director. The director is responsible for:

1. initiating a park;
2. creating the park, including its design and development; and
3. using and keeping the park.

The original focus of *Anatomy of a Park* was, and continues to be, that second item, the *creation* of the park. As the spokesman for the agency, and ultimately the person responsible for each park, the director is responsible for the design of a new park. Many park agencies employ their own design staff. But, whether in-house as an employee or by contract for a project, every director deals with selecting a design consultant for this work. At that point, the director becomes a *client*.

About the Consultant

Being a park design consultant is most often a service performed by landscape architects (LAs). The LA is a professional trained in all the issues covered previously in this book. Sometimes the service is offered by architects, planners, or engineers. Whatever the degree or professional title, the critical proof and privilege is only attained by achieving a record of successful parks. The record and the references are the most important considerations in evaluating a consultant for park design service.

Selecting a design consultant for a park project is essentially a search for a professional, the same as finding the right lawyer, dentist, or auto mechanic. You would seek out professionals with the depth of knowledge, the greatest experience, and excellent references from happy clients. Further, you would arrange to meet these people to be sure you can understand each other. And after making a commitment, you would treat this person with *respect* and work with him or her professionally. You would never lie to your lawyer, never bite your dentist, and never wait till a part falls off to take your car to your mechanic.

And never say "Green side up" to your landscape architect.

Maintaining a Good Relationship with the Consultant

Four key words—**knowledge**, **consistency**, **transparency**, and **flexibility**—represent the essentials that guide the client to the right consultant, and further keep the entire park development process headed the right direction to the satisfactory conclusion.

In the preceding chapters, we've introduced the *knowledge* of the essential elements and processes of park design that is key to the dialogue necessary in planning and designing a new park. This is the language used by landscape architects and park designers to explain the process of design as it progresses. The more familiar all parties are with the concepts and details of these elements, the smoother and faster the planning process proceeds. Time is reduced, budgets are reduced, everybody is happier. Good knowledge to have, especially at the initiation of Consultant Selection.

Knowledge is matched in importance by *consistency*. What is presented at the start of the consultant search should still be the reference at the conclusion of the search. For example, a project statement with scope of work should be maintained intact to the end of the selection process. If it must be changed, any change should be presented to all the candidates with ample time remaining for well-considered responses or revisions before the closing date.

Equally important, from the very first announcement of the park to the final approval of the finished project, the dissemination and sharing of information is fondly regarded as *transparency*. Neither client nor consultant will be pleased with the project if information is not freely shared. The ideal of "being on the same page" only works if everybody has all the pages.

Flexibility becomes important both to the client and the consultant in the typically fluid process of park development. Different employment procedures, funding crises, weather conditions, material shortages, political whimsy—all require calmness, imagination, wit: flexibility.

Alternatives for Employing Consultants

Consultants are retained for services in many ways. The most common are:

1. *A contract* for a specific park project; beginning to end.
2. *In-house design staff.* Sometimes this staff does all the design and construction documentation and project administration. More often, the professional staff becomes the contract administrator, selecting, advising, and guiding consultants from the design stage through the completed development. The value of

this situation is the department's LA constantly being available and involved, with the ability to advise on decisions: "That's not the way we do it."

3. *Continued or flexible service agreements*. Consultants are selected and employed, not on a project-specific basis but for a span of time, during which they will be assigned projects of various sizes and duration. These arrangements prove to be very useful for agencies with considerable activity and much variation in project size. A consultant may serve for a two-year span and work on 12 projects; others may have as much as a five-year contract and handle only three or four projects. This flexibility allows more productive time with less administrative time spent on contract preparation and legal processes.

Consider all the alternatives that are possible and be creative about new ideas that may fit special situations. The significant issue in choosing any of these alternatives is the potential savings of time and money, and accomplishing the best quality of work.

Finding Potential Consultants

Every project is an important project.

To the park agency it's the very reason for existence. Whether the next park project is simply an extension of the Pooch-n-Pal Dog Park or the development of a complete regional park with Olympic pools and a four-sport all-season complex, it's important for the agency as part of the continued growth of the agency.

For the consultant, every project offers the opportunity to do good work, leading to a continually stronger reputation—and to keep the cash-flow moving positively!

Depending on the current state of the economy, for the consultant, the dog park may be a much-anticipated opportunity to make a great impression on the local community and, not insignificantly, to keep the office staff intact and busy. For the regional sports complex it could be the gateway for a consultant to secure a national reputation for excellence. Consultants will be interested in considering every opportunity, and probably anxious to compete strongly to have the chance to secure the project.

Every park agency is a generator of work particularly suited to landscape architectural consultants. LAs are always interested in news of anticipated projects in the agency. To keep a positive link with potential design firms, consider some important opportunities:

1. *Be news*. Make sure the agency is well covered in regional news media. Make sure agency news is easily accessible on all the web-based electronic media sources.
2. *Read news*. Keep up with all sources of news about new parks and new types of parks. Who's designing these headline-makers? What designers are getting the most ink locally, regionally, state-wide, and nationally? Be especially aware of professional journals and magazines like *Landscape Architecture Magazine*, *Landscape Architect and Specifiers News*, and the American Society of Landscape Architects website.
3. *Get acquainted*. Maintain and expand lists of qualified, interested LAs. Make sure you know:
   - all the LAs in your immediate area;
   - most LAs within 100 miles;

- major firms in major cities 300–500 miles away (the basic criterion—how long it would take them to get to your projects).

Desirable major firms at greater distance can still be considered if they work with local firms on a co-venture basis (and they generally do). Be a "Speakers Bureau" to LAs and other park designers for their regional professional meetings. They *are* interested in you and your staff as well as your parks.

4. *Stay acquainted.* To maintain your consultant list, every year or two re-invite offices to update their portfolios in your files and encourage new contacts to submit portfolios as well. Equally important is your own review of what is in the professional file. On a regular basis, check to see if the firms that haven't been heard from in four or five years are still in the directories or on the Web. Conversely, don't let your reputation become that of the agency that always contacts the latest in the file (or, worse, the top of the pile).

Selecting the Right Consultant for the Project

From agencies across the country, the process of selecting a consultant has a few basic, consistent steps, which, by whatever local title, can be recognized as

1. the RFQ (request for qualifications),
2. the evaluation,
3. the short-list interviews, and
4. the selection.

The Request for Qualifications

Every public agency by law, custom, or common sense, makes a public announcement of the need for a consultant for service on a specific park or project. The first step in the selection process, this is presented as an invitation, called a Request For Qualifications (or RFQ).

The RFQ will likely *require* (with proof or references):

- experience on similar projects (including client references),
- the organization of the project team,
- the project manager/prime contact for life of project (with resume)
- the qualifications of team personnel and sub-consultants,
- the work plan and schedule for project, and
- the distance and travel time from the project and client.

The RFQ *might* request:

- communication procedures and resources;
- problem-solving procedures in meetings, including:
  — client/designer meetings, site team/contractors and
  — public dialogue with client, stakeholders, media;
- a description of project resources (e.g., personnel numbers, space available, equipment);
- a summary of current workload;
- design approach and philosophy;
- project management philosophy;
- an organizational chart;
- minority participation; and

- knowledge of and conformance with applicable codes and ordinances.

The firm may also be asked, "Will the person we interviewed be on this project throughout its development?"

The RFQ *should not* request the following items, which the consultant should view as an uncomfortable warning:

- a description of management strategy for a project with a restricted budget and a tight schedule (Is a project disaster being planned in advance?)
- litigation history over the last ___ years (regarding *suing* or *being* sued?)
- Corporate financial data/status (Why? What difference would it make?)
- A preliminary plan (or worse—*planS*) (Why? a gift bottle of Scotch would be cheaper.)

Note: If plans are specifically requested and a fixed fee is assured, this becomes a kind of competition. It also becomes a burden to evaluate, a misleading inspiration without adequate support, and a potentially serious embarrassment later.

The more "warning" kinds of questions or picayune, "labor-intensive-to-prepare" requests the agency includes, the greater the consultants' concern should be that the client is liable to obstruct rather than expedite the project. Agencies learn, with the loss of good consultants, that just because demands can be made on consultants under contract doesn't mean the consultant will ever come back to absorb more abuse, no matter what the next project.

The agency must make the maximum effort to provide complete, reliable framework information. A simple truth: the better the information provided by the client, the better the quality of the consultant's proposal.

Information a consultant *must* have:

- A base map or survey with topography and underground utility information;
- Project boundary information with restrictions in detail (e.g., access, utility ROW, adjacency issues);
- A program statement describing the intended elements and functions of the park;
- A budget;
- A target timetable; and
- Criteria to be used for the evaluation of the consultant's package and presentation, with priorities for each item.

Information a consultant *would like to* have:

- Community input regarding the program (an opportunity to attend open meetings or obtain records of such meetings);
- Design relationships, intentions, character;
- Preliminary plan studies, ideas viewed by the community (created by agency staff or by earlier contract);
- Important user groups for special attention; and
- Priority of program elements.

The Evaluation

After receiving the requested documentation from interested park designers, the agency reviews the submissions and ranks the candidates by criteria prepared and published as part of the RFQ documents. These criteria should be a part of the RFQ document so the consultants know the priority to give the parts of their responses. The fairest possible comparison will result.

From this ranking of qualified candidates, the list is reduced to a *short list* of most likely candidates.

Ordinarily a paper review of the RFQ responses would yield a ranking of respondents. The top 10 or 12 would be the short-list candidates. However, depending on the size of the project and number of candidate submissions received, a preliminary interview may be used to enable an evaluation of the top dozen (or fewer) very qualified candidates. Although these interviews would necessarily be compressed, they would allow a live impression of communication and personality qualities to help refine the short-list ranking.

The Short-List Interviews

Agency issues. The candidates identified for the short list are invited to personal interviews with a jury of representatives from the agency, during which the candidates will have the opportunity to present material about their firm. Additional documentation can also be requested and is also submitted by the candidates at that time, expanding the original RFQ material and providing greater detail about the firm's approach to the project. Typically scheduled for about an hour, the individual consultant's presentation is usually followed by a question-and-answer exchange with the agency.

When planning what the criteria for the evaluation of the interview will be, keep in mind that the qualifications of these short-list candidates have already been established. That's what the RFQ accomplished. Thus the criteria for the short-list examination should focus more on philosophical issues like design inspiration, the approach to planning the project, management strategy, and the phasing or timing of the various parts. This should be an opportunity for the agency to determine which of the candidates seems most inspired by the project and seems to have a larger vision of what it could be, no matter what the base. The interviews should also be a time for the agency to see the consultants lead (and read) people while expressing how they will work together to make the project better. The consultant who seems to be presenting separate viewpoints of his or her own personal strategies would seem to be less desirable than the consulting group speaking together about how they do work as a team, think together, and engage the client.

In many cases, an estimated fee may be included, usually sealed as a separate document. The reason for that is that the estimated fee should only be opened after the selection of the final candidate is announced. Thus the fee would not become a part of the selection criteria. The soundest selection is based on determining the best consultant, not the cheapest.

The candidates should be clearly apprised of the criteria and their relative value, on which their presentation will be evaluated. The jury should have a clear ballot with the same criteria and scores. After the selection is completed, the candidates should be informed of their relative position in the ranking of the candidates. How much information is released to the candidates is entirely up to the agency and will surely vary by agency and project.

Even though much is at stake in an interview of this kind, it is very important for both the candidates and the agency to be as comfortable and as open with each other as possible. This is a major commitment, and the two parties need to discover whether they can in fact work together comfortably or, conversely, whether they will collide as opponents. The most likely candidate is looking at the client representatives, hoping that they are willing to listen to process issues and recognize the need for program development as early as possible. The client representatives need to be alert and open-minded in the interview. The satisfaction that may come from hearing a candidate speak knowledgably about resolving details and making everything work is accompanied with the danger of thereby finding a successful problem solver who designs away from problems (thereby avoiding risk). The better target is the candidate who designs with imagination and discovers opportunity in the site's potential. Problem solving is important, but it's usually much easier than finding the fun in the project. All too often, the result, though less risky, is also less interesting.

As much time as possible should be allotted for questions and answers from both parties. Try to make sure all points have been covered, but also try to understand the consultant's thinking and level of creativity that will drive the process of design. Note the importance of making this short-list interview process as smooth as possible. The candidates should not have to confront each other when they are preparing their setup. Avoid adjacent waiting rooms if at all possible. Schedules should allow a few minutes for each candidate to set up his or her material. Ideally, a candidate group should be leaving the waiting room to set up their program right before the next group enters the empty waiting room—admittedly a difficult play, but worth a try!

The setting of the interview is entirely the client's responsibility. Although it's the candidate who is about to sell their company, the client is on trial as well. This will be the start of a long project experience. How good will the relationship be?

Make the setting comfortable for both parties. Keep the lighting abundant and balanced. Neither party will do well in bad light. Avoid an unbalanced windows-vs.-wall setting that places either party in silhouette. Reading faces is critical. Make sure to provide the opportunity for the candidate to stand, sit, or both as the presentation flows. No couches. And don't serve lunch! Somebody will spill it. Everybody will laugh (except the spiller who swears).

Consultant issues. If possible, find out where the interviews will be held and look at the space in advance. Determine what technology will be in place for sound system projection. Concentrate on your projects and your process. Client satisfaction is the most important selling point. Arrange to sit and face everyone in the client group. Avoid two rows for your party (so no one is behind anybody else).

With regard to the actual presentation:

- Don't hand out your slides on paper and then show them on screen! The jury will read what you just gave them and ignore you. Remember that illustrations are for viewing, not reading. Don't write your message on the screen and then read it. Always explain and enlarge what you're showing. The audience's imagination stops where the reading starts.
- Don't show up with free plans (read the RFQ section above—again). This is no different than any other bribe.

The Selection

Using established criteria, the agency again ranks the short list of candidates. The result is a meeting with the top candidate, during which the fee proposal and the proposed work are discussed and negotiated if necessary to fit within the agency's budget. If no settlement can be accomplished, the process repeats with the next firm on the list until a satisfactory conclusion is reached and a contract can be established.

You know this already, but everything must be fair and equal.

Before Anybody Steps in the Ring and the Bell Sounds. . .

The information in this chapter is not a set of rules or a statute of limitations. For every bit of advice included, there are folks with experiences very different from these, and highly successful practitioners who wouldn't do it this way. Their reasons fit their circumstances, as any guidelines will.

The point is to understand that this part of the creation of a park is an entirely people-to-people experience. Each party has to be sensitive to the goals and intentions of the other party. Successful relationships hinge on principles of fairness, sincerity, honesty, and integrity. The people involved have, and have a right to, pride in their professional work, This has to extend to an appreciation of the other professional's value and values.

The essence of the client-consultant relationship is a trading partnership. Each party trades their knowledge and resources with the other, all with the mutual goal of creating a park that benefits generations of people far beyond their time.

9 Good Park—Great Park

I thought I was a pretty lucky kid when I was growing up. Our house was right next to a park. Our street was called a "boulevard" (definitely not in the grand Parisian style). The street was actually two roadways, each about 20 feet wide, made of brick, separated by a 25-foot, curbed, grassed median. The park!

We lived on the corner. The boulevard terminated in a half circle at the intersection. Quite a few kids lived near that corner, so that end of the boulevard was the play magnet. We mostly played ball. It was whatever ball somebody brought, hit with whatever bat somebody could find. There were trees in front of most of the houses, but only one on the boulevard, in the circular end—and it wasn't much. It was an ailanthus ("Tree of Heaven," the wrongest name for the smelliest tree ever). Stubs of some of its adventitious children served as home plate. First base and third base were worn dirt patches at the curbs on the east and west sides. Second base was usually a hat. Since the street ran north-south, the orientation for the batter was excellent. A slide for first or third could get you a nose full of curb or a landing in the street where oncoming cars screeched their brakes a lot. Parents fussed about this.

But it was our park! We could fly little planes, We could attempt to fly kites. Anything we could think of. And . . .

That's what a park is for:

ACTIVITIES!

From the most intensely active, organized sports to the utterly inactive pleasure of laying motionless on the grass and looking at clouds, that's what a park is for. The activities can take on a planned status by becoming a program statement that lists and prioritizes the change of activity ideas into a physical form. The program gets built and Bingo! A park.

My park was a living program statement. But I'll admit it wasn't a good park. It didn't meet any size standards. The physical base was not very good. And safety? You could get killed there (and there were some close calls). So what is a good park?

Ask anybody what makes a great park and the answers would fill a book (other than this one) with opinions, some based on professional observations, but more probably based on personal recollection of things in the category of "I just felt good when I was in that park." Ask the same folks to name a great park and the answers would not even fill a long paragraph; most would be "Central Park." But the answer to the first question isn't the answer to the second question. No single solution makes any park a great park, or even necessarily a good park. The techniques for making a park are contained in the previous chapters, and the procedure for determining if a park plan apparently is properly prepared is also wrapped herein. But "good"? Or even "great"? Extracting a phrase from John Simonds, "Only the experience counts."

Pursuing this question of knowing a good park and, even better, a great park, I interviewed park professionals across the country. They were asked if they had, in their territory of concern, a great park. The answers were guided by their experiences interpreting the point of view of the users of the parks. The parks identified were all parks that people loved, that they used, that they would fight for if it were threatened (like seriously changing it or, worse, losing it). These were not parks the people used because they were the only ones available. They were parks that had characteristics and qualities that made them friends, that made them the fulcrum of memories.

Some park sites exist because they are places so strong in their original condition (great existing tree masses, incredible rock formations, water bodies overwhelming in their splendor) that even inept programming, planning, design, and construction couldn't screw them up. But good parks and great parks don't have to be tied to spectacular natural resources for their success.

From Ed Harvey, director of the Northbrook, Illinois, Park District, I learned that Northbrook's "great one" is Village Green Park. The place is near the center, the focal point of the city. It's got a gazebo, a bandstand, good athletic fields, and open space for casual sports. The best part (if you ask any of the kids) is the big play area in a grove of trees.

Figure 9-1
Cliff-beach: a spectacular natural resource. Nice, but necessary for a great park?

The core of what worked? Open space, shady places, strolling, fun people places.

Fran Beatty, former senior landscape architect for Boston Parks said Boston's best park—the "sacred one"—would be Boston Common. It too is located in the center of the city, with space for lots of people to be together, not just for programmed activities but also for unplanned pleasures. The park evolved from its original use in 1634 for cattle grazing, public strolling, and occasional hangings—all appropriate activities at the time requiring a large open space. The first real planning, over 100 years ago, was basically a competitive response to Central Park. One activity that continued to be popular from the original program was public strolling.

The water body called the "Frog Pond" has been redone for current ice-skating use. It's very popular for everybody, from "suits" down to little kids. Beacon Hill residents love it. Lots of other activities and elements have come and gone throughout the park's life.

Sound familiar? Open space, water, shady places, strolling, fun people places.

Bob Toalson, retired director of the Champaign, Illinois, Park District, didn't hesitate in identifying the great one as Hessel Park. Twenty-two acres of family-oriented, rolling ground, but in a gentle bowl form lending an immediate central space. It's also the most popular picnic target in the region. Why? From its origin in the 1920s as a strolling, open-space park, the early efforts of the Civilian Conservation Corps programs produced a mature grove of tall shade trees complementing the natural space-forming potential of the landform itself. Peripheral parking has maintained the space with no through-circulating traffic. From its original "keep off the grass" image, Hessel Park has evolved into a completely open space, with various play courts and full, very intense use of all areas.

Open space, shady areas, strolling, fun people places.

Claude Thompson, former head landscape architect for Dallas City Parks and currently planning director for the city of Wylie, Texas, commented on two very big (Texas big), very popular parks with similar characteristics. Both White Rock Lake Park and Rochester Park have perimeter drives that feed into cul-de-sac or dead-end parking sites serving picnic and play areas. Large water bodies are located in the center of the parks, and huge open, uncommitted areas are scattered through both. Plantings tend to be around perimeters, where they don't interfere with the large open spaces, and they accommodate the picnic activities in shade as well as provide a sense of personal scale. Landforms? Ain't none. (Texas, remember?)

The huge open spaces are very popular for large group activities, reunions, un-programmed play, and spontaneous stuff like unorganized games. Drives and trails allow ample space and locations for running, skating, and non-vehicular travel around the parks.

Open space, water, shady areas, strolling, fun people places.

You'll find all those same things on a huge scale in Grant Park along the Lake Michigan shoreline in Chicago. Although Grant Park is virtually the park symbol for Chicago, Bob Megquier, retired head landscape architect for the Chicago Park District, pointed out the more significant people impact of the many neighborhood parks distributed throughout the city. The century-and-a-half-old history of Chicago parks includes their importance as open-space resources for unprogrammed uses (including their role as a haven for refugees from the Great Chicago Fire of 1871). Originally focused on field houses as central service and entertainment points with adjacent space for field sports and strolling, the parks have continued to be flexible in the use of those same spaces with the transition of the field houses into recreation centers. The parks of Chicago, many of which have the design signature of Jens Jensen in their history or the Olmsted patent so widely distributed across the country, constitute a great example of a collection of parks evenly knit into the fabric of a city. While the premier place, Grant Park, may be the focus of the "occasional special" (e.g., the Jazz Festival or a major holiday celebration), the neighborhood parks host special occasions that keep the neighborhoods alive: family picnics, reunions, church socials, local team sports, and school parties.

Before going further, return to the experiment from Chapter 4 using Google Earth. Access Google Earth and locate your favorite park, the place you know well. Zoom in to a view of the whole park. Looking at the aerial view (the live plan of the place), try walking the paths, driving the roads or lanes, sitting in the spaces, exploring the whole place. Think about what seems satisfying, successful, enjoyable. Think also about elements that are the opposite of satisfying, successful, enjoyable. Does your park include open space, water, shady areas, strolling, fun people places?

If not, are there changes you can imagine that would bring more of these good things to your park? Give your park a rating—adequate, good, or great. You're improving your critical park-appraisal ability by using this exercise of overviewing as well as through confirming your impressions on the ground. Keep practicing.

Predicting the Future by Design?

Briefly—we can't. If this volume conveys nothing else, surely the importance of communication between all parties involved in park development is clear. The new ideas for the future won't originate with the designer. The designer can expand and give form and detail to the client's ideas. But ideas start with clients. And they *are* the future.

Anatomy of a Park continues to stress the importance of the relationship between client and designer. This is not a chicken-versus-egg question as far as which comes first. Frederick Law Olmsted, for his first project as a landscape architect, responded to the client, the citizens of New York, to win the competition for the final design of Central Park. His entry in the competition responded to an idea in spirit and gave form to what then became the program for the ultimate design.

In the last two decades, the growing popularity of skateboards looked like a program for the future and has resulted in skateparks added to parks of all sizes. User needs for these facilities were expressed to cities and their park agencies all over the United States. And park agencies responded with skateboarding facilities. But these weren't total-park facilities. They were adjunct specialties. Based on ad frequency, interest seems to have waned. The youngest don't have the visible hero figures who once appeared in the sport. However, like snowboarding, skateboarding is evidently aggressively on its way to becoming an official Olympic sport. Skateboarding facilities may surge again.

But at the end or peak of this cycle, a basic question: Are the parks better? Has a great one emerged from this cycle? No. The parks making these additions have proven their flexibility and responsiveness to their populations. But no great parks were created by the addition.

The Envelope, Please!

Will parks, guided by societal swings, with well-thought, well-designed answers to all the programmed needs, turn out to be true memory makers? What challenge will produce the next competitor for Central Park? Olmsted's creation established a very high bar for future candidate parks.

A significant possibility has been created in this decade, again in New York City. The park, called The High Line, was born as an idea from a group of citizens who received with great concern the announcement that a major elevated railway structure, having been abandoned, would be torn down. The rail route ran a fair distance through central sections of the city, with powerful views into its heart and out to the river. Their concern was shared and multiplied by a large group that organized itself around the question "Why do you think the High Line should be saved?"

Bolstered by the depth of concern for the loss of the striking structure itself and the scars it would leave, the citizens took the positive view of "What Could Be Made of The High Line." From a long series of very productive meetings, the group (now definitely The Client) realized that all the great ideas needed The Designer.

To make this next huge step, first the High Line Foundation was created. This was followed by the announcement of an open competition to identify designers who would have the talent to become the design leader of the High Line project. All the details of this entire process and the competition, the selection, the design process, and video tours of the astounding result can be found on the web simply by Googling "The High Line."

The result of the competition was the selection of the James Corner Field Operations Office as the design lead. From its inception, the project has been truly unique. The site (the railway) is 1.45 miles long and is about 50 feet in width! Of course there are linear parks longer than this on abandoned rail sites, but not this high aloft.

The design creates a front yard for people several stories in the air. It opens "windows" on long views far across the city and out onto the ships harbored opposite. The client is an audience of all ages, capacities, and interests. The design program has created subprograms of entertainments through dozens of different environments—architectural as well as horticultural.

Open space, water, shady areas, strolling, fun people places? Right! Got it all. Looks like another great park has been created.

There is no way to guess what's next. We can learn from the experiences of previous park officials confirming that, just as there are basic design characteristics that prevail in a successful park, there is a consistency in how user needs are satisfied in park design, no matter how fantastic they may seem at first. The park system needs to be kept as flexible as possible to meet this future challenge.

The Park System

All the parks in the various locations we've mentioned reinforce the fact that, no matter where a park is located, it is unlikely that all people's recreation needs can be met on one site in one park. A number of parks are usually required. This collection of sites serving the collective need is the *park system*. Its planning involves four steps:

1. At the outset, an inventory is made of existing parklands, including acreage, location, and activities provided for. This shows where things currently stand.
2. To determine what there is to work with, existing acreage is compared against national standards considered to be the minimum for the population served. For instance, such standards suggest that a small city or town should have 10 acres of parkland for every 1000 people. Thus, a raw goal is established.
3. The next step is to determine where in the district the land with the greatest recreation potential is situated. These parcels are identified through analyses of existing natural and cultural conditions. Then studies concerning population concentrations, future growth directions, and leisure-time interests of affected age, sex, ethnic, and income groups are conducted to provide information that will enable planners to ascertain which pieces of the high-potential land should actually be acquired.
4. Finally, acquisition priorities are spelled out and avenues of financing are investigated.

The product of these steps is the *park system plan* (sometimes called *comprehensive plan*), offered as a map showing the location of both the existing and proposed park sites. (See Figure 9-2 for an example.) Once purchased, each parcel must be given further study to chart its course of development, but many decisions affecting the success of that development have already been made at the stage of acquisition. If the land is not suited to the activities desired, either program or site will have to be compromised. That is, program units will have to be eliminated because of site restrictions, or the site will have to be physically transformed in order to shove in nonconforming facilities. Park system planning deserves a volume in itself; we cannot even attempt to scratch the surface of the subject here. However, the following recommendations are well within the scope of this brief introduction to design. An expert with a feeling for design and knowledge of development should be a part of the planning team. Ideally, desired uses should be identified before sites are acquired. If this is not possible, *variety* should be a prime criterion ordering the purchase of land. If sites exhibiting a wide range of characteristics are available, there is sure to be a parcel suited to a future need.

When that need arises, design begins. Will it be a quality solution? Remember, you will have to live with the results.

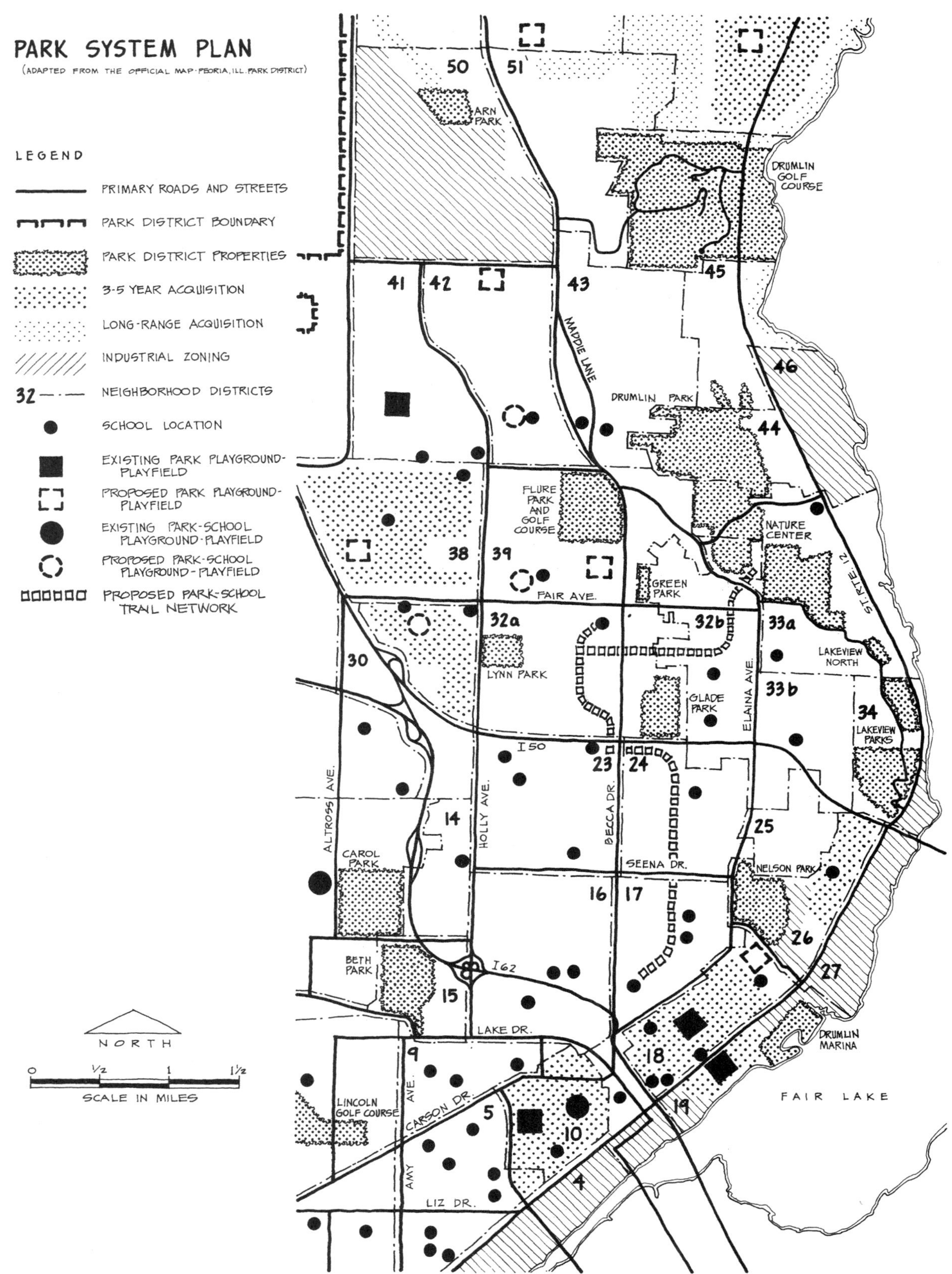

Figure 9-2
Park system plan (adapted from the official map, Peoria, IL Park District).

10 The Bottom Line

As discussed in previous chapters, the park board, with the help of the park staff, citizens, park users, and maybe a consultant, must develop a program for park and recreation facilities. The program may be for one park or for many. It may be simple or complex. The vision may be wildly grandiose or mildly frugal. However the plan is assembled, it is only a plan until the park is built. The actual process of building the park immediately brings money into focus and introduces the issue of budget.

Developing a budget is usually an educational experience, but it is often an unexpectedly painful exercise as well. By the time you begin to look at a budget, a realistic assessment of needs and serious discussions of priorities must already be behind you. Eventually the prioritized needs and the budget must come together in order to build a new park, make park improvements, or simply maintain the status quo of the existing park system. It is necessary for a park board to compare the estimated costs of park improvements and maintenance to the available budget as declared by the city council or determined by tax revenues.

It is important to consider a variety of ways to achieve the program needs and simultaneously satisfy the needs of the community. Having lots of money is a simple way. But even without cash, much progress can be achieved. There are other ways to get parks built or assist in cost control without cash. There are also ways to make sure you get the most for your money. Leveraging available budget resources can achieve results that otherwise may not be possible. The term *leverage* is similar to its physics concept: an amount of money or other commodity of value is used to achieve or obtain a greater amount of money or valuable assets.

The following budget-expanding tools and techniques are arranged in cash (i.e., grants) and non-cash (i.e., volunteers) ways of achieving park goals. The assumption here is that the park board starts with a fixed budget from its normal appropriations and searches for ways to grow that budget to achieve its vision for the park system. The time frame for park development can change to accommodate the budget and plans, but if the vision is unclear, no amount of time and money will bring about intended results.

For the sake of discussion, let's say the park board has a well-defined budget derived solely from city appropriations. They have just completed a five-year master plan with enthusiastic input from numerous citizens and careful cost analysis from the park staff. The budget covers all but a few high-cost construction items. Priorities could be arranged to distribute these items over the five-year span. The budget needs to be stretched—more money.

Cash-Expanding Ways to Grow a Park Budget

The Big Bucks—Sell the Park! (But Keep It)

Imagine a major dream come true—a grand, regional park with the potential to attract visitors from a huge area as well as serve the city with the ultimate menu of facilities for an excellent recreation program. Or how about a sports stadium/fields venue for tremendous league play of every kind, with the additional opportunity of visiting concerts, musical shows, circus performances—the works!

The only obstacle to achieving this dream is a cost well beyond any tax levy or fee structure imposed on the city's economic base could bear. No hope? That depends on the crowd-drawing capacity of the actual development. If the city is happy to own the park, and if the developer owns its name, a serendipitous situation can result. Corporations are willing to invest their advertising money in a project that produces advertising power beyond other outlets. Consider: Los Angeles's Staples Center—LA Lakers; U.S. Cellular Field—Chicago White Sox; Minute Maid Park—Houston Astros; KFC Yum Center—University of Louisville, men's basketball, and major shows; State Farm Center (the original, newly remodeled Assembly Hall)—University of Illinois basketball plus major shows.

What's in a name? Possibly a facility beyond an agency's dreams. Just as there are design and planning consultants, there are consultants for the financial considerations, the staging of development, and the facility's programming and revenue stream for such facilities. Before a project of such a scale is launched, wisdom suggests identifying and engaging a design team with the background to take such a project from the idea phase to active implementation. In short, once an agency has worked out how to own its own lion, no matter what its name, there also must be talent capable of keeping it fed.

Grants

On a more modest scale, grants may be the most obvious way to expand a budget. Grants allow the park board to leverage money in their budget to fund projects that they otherwise could not afford. A grant provides money from an external source. That source must have enough interest in your project to reward your application with money. Granting agencies may be federal, state, or local government agencies or authorities, not-for-profit or civic organizations, or foundations.

There are several steps in the grant request/award process (from the granting agencies' perspective):

1. *Source of money.* Yes, what you may view as a source of grant money must also have its own source of money. The source of money may be the state legislature in the case of state agency grants, or it may be interest on large sums of money in a family foundation's trust fund.
2. *Grant administrators.* These are the people in charge of deciding how their money should be distributed. Grant administrators can be individuals or groups, directed by an agency head or by a board of directors. They have a mandate to award grants to projects that best fit their agency's mission and giving goals. The grant administrators draft rules regarding the grant recipient's eligibility and the project's suitability to their mission. They decide on the size of awards to be offered and the requirements for a match (defined later). They later review the grant applications to determine which ones get funded. It is important to remember that the grant administrator is paid to award the money to worthy projects. If your project makes them look good and makes their board look good, everybody wins.
3. *The grant application.* This is the beast that many people fear. Park boards shy away from grants because they appear complicated. (Some come with inch-thick instruction manuals.) Park staff members are quite sure that they don't have time to write a grant application on top of all their other responsibilities, and park design consultants just want to design parks. In reality, any of them could do it. Park board members have written successful grant applications and lived to tell about it, with pride. Park staff have written grant applications and found themselves in favorable positions for promotion and pay increases. And many design consultants have discovered that writing grant applications is another service that they can offer.
4. *Project implementation.* Once the grant is awarded, the project goes forward and the park gets built. The project, however, must get implemented according to the rules, regulations and constraints of the original guidelines of the grant program. This is the essence of the entire process; if this procedure isn't followed, none of the rest of the process counts.

Grant funds, because they are predominantly from public sources, usually require a local match. This is an amount of local money or other in-kind contribution to the overall project funding. For example, if a community park board decides to apply for a state grant, one of the requirements likely will be that the community pays a percentage of the project costs. This amount may range from 10 to 90 percent of all costs. A grant that covers 90 percent of project costs is probably worth going for; one that requires a 50 percent match requires more scrutiny; and one that pays 10 percent of the cost is questionable. Although the local match may have to be cash, in many instances it may be satisfied with in-kind contributions of labor, equipment, construction materials, or the value of land purchased or donated. This helps a cash-poor community compete for the available funds. They may have more volunteer labor hours than dollars to offer. The granting agency decides. It tries to determine the extent of need for the types of projects they want to encourage. Then it decides how competitive to make the grant criteria. Finally, the grantor writes the guidelines for the use of the funds.

Grant guidelines are often referred to as hoops, hurdles, or strings attached. They're the confirmation of the old quote, "There's no such thing as a free lunch." Broad guidelines may include any kind of park development. Narrow guidelines may be targeted for specific kinds of activities, such as trails or nature areas. Everything depends on the priorities of the granting agency or its interpretation of the priorities set by the entity that appropriated the money.

Let's say the regional council has appropriated the granting agency the sum of $1 million to enhance the parks in a five-county area. The agency could give $200,000 to each county and hope the money gets spent in the right places. But the director decides to announce a grant program in hopes of attracting proposals for the best and most needed improvements. She asks a grants coordinator to write two announcements for review by the board, and he gets right on it.

Scenario 1: The grant is written for a broad range of eligible projects. The maximum dollar amount is $200,000 (good money) with a match requirement of 20 percent (cheap money), with a broad definition of in-kind match (easy money). Everybody is going to chase this grant. The result is that the agency gets 22 applications totaling $4 million for the available $1 million in grant funds. The agency has to make some tough choices.

Scenario 2: The grant is written for paved nature trails (narrow focus). The maximum dollar amount is $100,000 (not bad money) with a match requirement of 50 percent (not so attractive), with a required cash match (difficult). Many park boards are going to think twice before applying. First, they may prefer unpaved nature trails. To get the money, they must construct a project in a way they do not like. Also, the amount of the grant is not enough to finance the entire project. Finally, they could do the project with the contribution of available trail-building materials and volunteer labor but would find it very difficult to come up with a cash match. There may be a shortage of applications for this grant program.

Remember that the agency's job is to allocate the money. If they write guidelines that are too restrictive or not restrictive enough, they have a problem. The agency will adjust its criteria until the funds available approximately meet the demand.

True Story: An unnamed national conservation organization decided to offer a grant program to promote its ideals. The organization offered an attractive amount of money (for activities, not construction). It offered rather vague guidelines for the purpose of attracting a variety of project types. This was a grant with no match requirement. The granting agency received over two thousand proposals for five awards. It takes a huge amount of time to read that many proposals, even if they are only two pages each. Two or three people typically read the grant proposals so that the decisions don't fall on one person. Do you see the problem here?

Returning to the work of the grantee, writing a grant application or proposal is not an act of magic but rather a studied, logical, and practical process that anyone can learn. The most difficult challenge to overcome is the fear of doing something new. Following are the five basic steps to writing a grant application:

1. *Request and/or receive a grant application package (forms and instructions).* First-time grant writers may have to request an

application package from the granting agency. Thereafter, you may be on their mailing list and get announcements of new or recurrent grant opportunities.

2. *Read the application and instructions*. An application form may be only a page or two in length, but the instructions for completing the form and describing the guidelines may be a hundred pages long. The guidelines contain several very important items of information:

   - *Project eligibility*—The project for which you are seeking funds must match the stated purpose of the granting agency. If the agency is soliciting applications for trail projects, you will waste your time asking for funds for pool construction. The project eligibility may be called something else, but it will be quite easy to compare your project with the project types being funded.
   - *Recipient eligibility*—It is usually quite clear in the application form which kinds of recipients are eligible to receive their funds. If the agency requires that the recipient be a park board, a not-for-profit organization need not apply. If the recipient is required to be a duly ordained park board with an approved five-year master plan, a simple park committee may not be eligible. Often the first step in pursuing a grant may be the approval of a master plan. Read and reread the application guidelines and restrictions. If any part is unclear, call the granting agency, which will be eager to help you interpret the guidelines. The agency's job is to put its money into the projects that best meet its mission.
   - *Available money/maximum request*—It"s important to know how much money is available from a given grant program and the maximum per recipient. Since most grant writers will request an amount close to or at the maximum, dividing the total amount by the maximum (per recipient) will yield the number of grants to be awarded. This also is a good indicator of your chances of being awarded a grant. The more grants there are, the less competition there is. A small number of grants creates a high level of competition. The grant applicant must have the ability to assess the competition and the chances of getting an award to determine whether or not to submit the application. The grant-application process costs money either in staff or consultant time. If there is little chance of getting the grant, there is no reason to apply. Wait for a better chance. There is, however, one good reason to prepare a grant application even if there is a slim chance of getting the award. The application will be written, everyone has a chance to review it, and it will essentially be ready to be submitted for the next grant round or might be adaptable to another grant program. An honest try may also engender some very helpful feedback from the grant coordinator.
   - *Match requirements*—Some grant programs offer free money, but most require a *local match*—a prerequisite demonstration of willing participation on the part of the local community to put their share of money into the proposed project. The logic of the granting agency (grantor) is that if the community is willing to put some of their own money into the project, then the grantor will consider becoming a partner in

the project. The local match may be any percentage from zero to a hundred (although a 100 percent local match would cause one to wonder what the grantor has to contribute). A short story: One of the authors has taught a grant writing class many times. He usually introduces the basic concepts of grant writing and then challenges the students to write a lunch grant. They have to figure out how to feed the class with a grant of up to 50 percent from the professor. Their match is an agreed-upon "revenue" of two or three dollars per student. (The local match must be 50 percent or more.) For a class of eight to ten students, the professor has had to fork over anywhere from zero to thirty bucks. Yes, on occasion, the grantor has gotten a free meal, but then he did provide the incentive (assignment). The class ate well and learned the finer points of grant writing.

- *Other requirements*—All other requirements can be combined at this point to indicate those hoops and hurdles referred to earlier. These are usually clearly outlined requirements that relate to the purpose of the grant program and the mission of the granting agency. Only an overly zealous grant-writing consultant will overlook or misinterpret these requirements to erroneously try to squeeze a project into a grant program that it is not meant to fit. There is no reason to write the grant if the project cannot satisfy the requirements. If the requirements inflate the cost of the project, getting the grant just might not make any sense. Look for grant administration, accounting, and reporting requirements.
- *Due date and cycle*—Most grants specify a very specific due date and time of day. Grant deadlines are probably as inflexible as any you will ever encounter—less flexible than your average client or college professor. Some grants are reviewed and awarded on a rolling time line (submit at any time), but these are exceptions rather than the norm. Many grants have quarterly or semiannual deadlines. Some grants cycle every year or two. This information is clearly stated.

3. *Call or contact the grant administrator or coordinator.* After reading the application, you surely will have questions. Call or e-mail the grant administrator to get those questions answered. Even if everything is explained to your satisfaction, contact the administrator to discuss your project and your organization. The administrator likely will be eager to learn about your project and others to estimate the number of grants being submitted. The grant coordinator can give you an early indication of your chances of getting the grant.
4. *Prepare the grant application*. Many grant application packages will be accompanied by an outline or checklist of required information. Use it. This will be the best assistance you get in preparing the application. You may have a better idea of how to organize your presentation, but don't use it. Your job is to make the grant reviewer's job easy. The grant coordinator has provided the checklist for his or her own convenience; don't confuse the issue. The application may seem to request redundancy. Go with it. It is better to be redundant than to leave something out. Start with a synoptic narrative description of your project. Use this project overview to answer the questions

or provide the details requested in the application. Get feedback from your client, fellow staff members, or the park board as you proceed. This is the best way to earn and maintain support for the project and grant application.

5. *Wait.*

   Now is not the time to call the grant reviewers. They are busy reviewing grant applications. Don't send addenda or corrections. You had your chance. You should receive a note of receipt. Better yet, send it by certified mail with a return receipt or by another traceable courier.

   Prepare to receive the grant by ensuring that you are ready to do the requisite accounting for the project. Get your ducks in a row, whatever that means.

Sponsorships

Sponsorships are another way to extend the budget. Sponsorships are negotiated with corporations, foundations, and local companies. You've seen "sponsored by" signs on the tees of some municipal golf courses. Those are not just friendly little billboards. They are a serious means of bolstering the budget of the golf course and perhaps the entire park budget.

Unlike the grant program, a sponsorship is less likely to provide funding for an entire project. Sponsorships are commonly negotiated for a golf hole, a playground, or a section of trail. While granting agencies often require a sign of credit, sponsorships usually require named facilities. It's the Bill Jones Ford fifth tee, or the Caterpillar Soccer Park.

Sponsorships are by invitation, but the park board and staff must know what to ask for and whom to ask. For instance, if the parks department plans to operate an eighteen-hole golf course with reasonable user fees, it may be necessary to enhance the income side of the budget with hole sponsors. Depending on the operating costs, a sponsorship may cost very little. If, however, the plan is to renovate the existing nine and construct an additional nine, the cost to hole sponsors could be quite high.

It is important to have a plan with a realistic cost estimate. This cost data can be entered into the budget, along with other known costs and project revenues. The budget shortfall divided by the number of holes would be the approximate average cost of hole sponsorship. A hole sponsorship is preferably a yearly cost with a five- to ten-year commitment. Why make new signs every year?

When the numbers are in and the cost is established, go to potential sponsors individually or as a group and ask for their support. Before you go to the group, test the water. There may be a corporation interested in sponsoring the entire project (golf course or other facility). One is simpler, but several sponsors ensure more buy-in across the community.

Sponsorships are not necessarily gifts. Corporate gift-giving allocations may be small, but a sponsorship is an opportunity to "buy" advertising space on a sign or in a brochure. Advertising budgets are usually quite large. That is the pot to tap. The excellent exposure is of value to them; that's what you have to sell.

Non-Cash, Budget-Expanding Ways to Get Parks Built

It may be exciting to contemplate ways to collect or generate cash for parks, but there are other ways to accumulate material or services of great value. The following techniques may not generate cash, but they may reduce the need for cash. Do you want a grant to buy a new piece of playground equipment or a new piece of service equipment at no cost? Either way, you get what you wanted. Non-cash, budget-expanding methods can be put into three categories: partnerships, technical assistance, and volunteerism.

Partnerships

Partnerships are born when two or more entities (such as a park board, local business, civic organization, or scout troop) discover a common interest in park improvements or changes in park programming. The partnership grows when the partners learn that they can better achieve their goals when they work together than any of them can individually. "Friends of Parks" and park boards are natural partners. The park board is eligible for certain government grants that may not be available to the "friends" group. The Friends of the Park, a not-for-profit corporation, may be eligible for grants that are not available to the park board. The Friends can buy and hold lands that can be used later as a match for a park board grant and may be able to conduct fund-raising events that the park staff could not handle. Friends can also spend money without being subjected to the sometimes cumbersome government procurement procedures.

Partnership opportunities may be "too close" to see. Maybe the park board can't see the urban forest for the trees. Perhaps the urban forester is in the street or utilities department. A partnership, which is already legally intact, can develop between city departments. If the street department has an urban forester who knows how to get money (or has money) to plant trees, solicit her help in getting trees for parks. The park board has to have something to offer in return. Partnerships are not built on requests alone but rather on trading goods, sharing responsibilities, and celebrating successes.

A growing soccer league may be able to raise the money to prepare the site, install a parking area (another partner may be a crushed-stone quarry or a paving company), buy the goals, and staff the soccer program. And during the time when the soccer league isn't using the facility, the fields are open to the public.

Another example of creative partnering is one between a school board and park board. The school had an outdoor pool for its team that was grossly underutilized during the summer months. Kids wondered why they couldn't go swimming in the unused pool. The park and school boards got together and crafted a legal arrangement so that the pool literally (legally) becomes park property after 3 PM during the school year and during the summer break. Two problems were solved, operating costs were shared, and everyone was happy.

Trail projects offer a variety of opportunities for partnering. Trail partnerships may form between adjacent towns and counties, not-for-profit corporations and schools, and other obviously or tenuously related entities. The partnerships may combine to solve problems of land acquisition, trail development, or trail maintenance and security. Trails have provided good examples of partnerships between very different entities. Police agencies may partner with scouts and a trail advocacy group to provide security while the same advocacy group, a local hospital, and a local sporting goods store develop a partnership to promote the health benefits of walking and conduct periodic fund-raising events for trail maintenance.

Not every partnership must be for life. It's not a marriage. A partnership is created for the length of its usefulness. If a partnership is mutually beneficial to the partners for only one week, one season, or three years, it can be dissolved when the its goals have been achieved. Furthermore, a park board can develop multiple partnerships for different goals or projects, and each can be on different terms and time frames.

Technical Assistance

Technical assistance could be offered as part of a partnership, but it can also be a stand-alone offer that may be too good to refuse. You can write a grant to purchase technical assistance or find free technical assistance, which may be offered from a number of different entities such as government agencies, corporations, national or state not-for-profits, universities, or knowledgeable individuals.

Government agencies may have the technical assistance you need. The agencies that offer grants will gladly provide technical assistance for preparing a grant application, but that is not the subject of this section. A park board may need real technical assistance regarding real problems and issues, technical assistance that has value. State departments of natural resources, environmental management or protection, transportation, or public safety offer technical assistance for a wide variety of situations. Staff in these departments often have expertise equal to or greater than their counterparts in private practice. They consult with many communities and park agencies. They have a broad perspective of the problems and associated solutions that have been tried and succeeded in many different locations. They are specialists and experts in their fields.

The National Park Service has a River, Trails, and Conservation Assistance (RTCA) program, for example, that provides technical assistance in a variety of projects, as the name implies. Their staff serves a given geographic area and freely assists communities (park boards and others) in the transfer of ideas and techniques from one site to another. They also attend, sponsor, and participate in state, regional, and national conferences to increase their own knowledge and to further prepare and equip themselves with the information that they will share with the communities they serve. So when someone from the National Park Service comes to your park board meeting and says, "I'm with the government and I'm here to help," funny as it may sound, welcome them to your partnership.

There are numerous examples of technical assistance offered by other government agencies, but suffice it to say that many other possibilities exist and that the way to find assistance is to ask for it. Agencies often create partnerships to develop grant programs or technical assistance programs because they, too, recognize that not all problems are single faceted. Their programs are tailored to meet the needs of the communities in their service areas. Appendix 6 lists some government agencies offering technical assistance to communities.

Corporations may offer technical assistance. They are a better source for sponsorships and gifts, but depending on the business they are in, they can also offer valuable assistance. Corporations usually have a variety of departments with various services to offer. They may have a legal department willing to provide the parks and/or recreation department some free legal services. If you are intending to establish a not-for-profit Friends group, legal service fees can accumulate quickly. Corporate technical assistance can be extremely valuable. The marketing department may offer assistance in promoting a fundraising event; the graphics department may be able to make signs for a trailhead or print banners for an event.

Corporations also may be able to offer space in their suite of offices to your Friends group. This is not technical assistance, but the proximity to people with valuable skills is an opportunity to learn by osmosis the knowledge that will further the mission of the Friends and the parks department.

This kind of assistance is not money coming in; it is money not going out. Money not used for donated technical assistance can be used for other expenses. This is not rocket science—it is just thinking creatively and watching diligently for the right opportunities.

Not-for-profit organizations offer technical assistance that is as good as cash. While technical assistance from government agencies may tend to be broader in scope and freely involve more than one agency, not-for-profit organizations will more likely want an exclusive connection with your park board or community. Not-for-profits are more competitive for the attention and the membership support of the local community and the larger community of supporters. They want people in the community to know that they provided a valuable service, and community members are then invited to join their organization (give money) and continue their mission.

Nevertheless, the technical assistance of not-for-profits is very valuable. For example, the National Trust for Historic Preservation (NTHP) provides an excellent service for communities that may have historic structures in need of repair or restoration. So if the park has, for instance, a WPA-era park shelter in need of repair or stabilization, NTHP can help find the resources, local or elsewhere, to fix the structure.

A land protection organization may be interested in preserving land of exceptional ecological and aesthetic value. The technical assistance they would bring to the table would be closely related to land preservation and protection of ecologically sensitive areas. So, while they may provide excellent assistance in the protection of the land through the purchase of conservation easements, they may not be the best source of information for the latest and greatest paving surfaces for trails. However, they may be quite interested in an ecologically sensitive interpretive trail and boardwalk around and through a special wetland.

A not-for-profit's technical assistance may not always be free. They may require an application for services. There may be competition for their assistance, or they may deliver the assistance in the form of a grant from a foundation that requires a match. Some organizations may offer assistance for establishing a local not-for-profit for the long-term management of a project. There are many ways that technical assistance can be offered. The parks department must be sharp enough to recognize opportunities when they appear. Technical assistance is not worth asking for if it is ignored, but if it is requested for a reasoned purpose, delivered by a knowledgeable staff, and implemented with wisdom and enthusiasm, the results will be a huge success.

Consider the source of technical assistance. The park board and staff should have a clearly defined mission; so does the not-for-profit.

If the two are similar or compatible, ask for their guidance; if not, move on. There are other organizations that undoubtedly have an interest in your project and would be glad to help.

Community foundations are closer to home. These are not-for-profit corporations established in communities for the specific purpose of funding projects that may not have access to other funds. The staff of a foundation is probably very astute at fund-raising, since they must raise funds in order to distribute them. If your project does not qualify for their funding, they may be of great assistance in identifying other sources. Appendix 6 also lists not-for-profit organizations offering technical assistance to communities.

Universities offer a number of options for technical assistance. All universities have some combination of missions in research and discovery, teaching and learning, and outreach or engagement. Communities and park boards can benefit from one or all of these. Certainly research in recreation and leisure studies, environmental psychology, and resource management, to name a few, are important to park departments. Often the state recreation association and the university will collaborate to conduct annual meetings at which the latest research findings are reported. Park personnel will find these conferences critical to staying current with developments in the field.

Programs in landscape architecture offer quite a different brand of technical assistance in the form of community assistance or service learning projects. Actually, there are many other park-related programs that offer various kinds of technical assistance (but since the author is a landscape architect, that is where the focus will be). Of the many graduate and undergraduate programs in landscape architecture, most offer some form of assistance to communities, some of which is for parks and recreation.

The assistance comes typically in the form of feasibility studies, visioning, or master planning. The product of the students' work is rarely a set of construction documents, but the product inevitably helps the park department move forward. The projects may relate to the protection of natural areas, an evaluation of the park system, an inventory of trees in the parks, or the development of trails and greenways.

The benefits of these projects are really threefold. The students are challenged with real-world, hands-on projects with real clients. The community benefits from the ideas generated by the students under the supervision of a knowledgeable instructor who probably has practical experience as a park designer. The profession of landscape architecture and park planning benefits by the increased visibility. Additionally, the students become involved in the community and provide a voluntary community service, as do park board members.

The university usually charges a small fee to cover the administration and expenses (such as travel and production) of the project. These projects are not meant to replace the services of planning and design professionals but rather to assist the community in getting started. Often the community has no collective vision; they cannot see the way to improved park facilities. Getting some ideas down on paper is all that is needed to jump-start the park board and staff.

Perhaps the best assistance from government agencies or universities is information about grants, sponsorships, and partnerships that might help the community to leverage their money. Leverage, as mentioned earlier, is using a small amount of money to acquire more money or goods and services of value. Whether the leverage is in the form of cash, donated equipment or services, or the provision of tech-

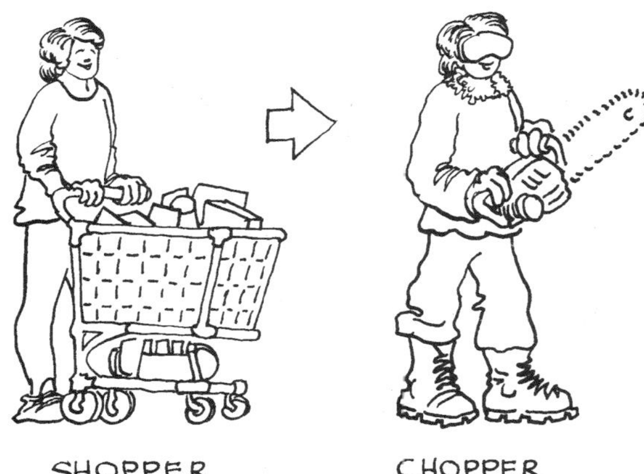

Figure 10-1A

nical assistance, the main idea is to stretch the available resources. Appendix 6 includes a list of universities offering technical assistance to communities.

Knowledgeable Individuals

Knowledgeable individuals come in many shapes and sizes, and they possess all kinds of valuable information and expertise. Many of them are retirees from government agencies, corporations, and national or state not-for-profits and universities. These individuals may be self-employed or without need of employment. Only you know who these people are and how they can help you. They may not know that you need them.

Volunteers

Volunteers are as good as gold and represent yet another way to extend the capacity for getting parks built and maintained. Volunteers are everywhere. (See Figures 10-1A and 10-1B.) Everyone is a potential volunteer waiting to be asked to contribute to a meaningful activity or project. People who volunteer are attracted to tasks that are fun and give them an opportunity to socialize with others of similar interests.

Park board members are volunteers who give their time, energy, and expertise. They put their names and reputations on the line. Their decisions are in the news. Their mistakes are on the front page, and their successes are often hidden near the legal notices. The public, through its election of park board members, entrusts its money to their judgment. The park board members have significant budgets to manage. They must ask for and consider the requests, ideas, and suggestions of the public and then draw on their collective wisdom to make the right choices.

On a more positive note, the park board reaps the rewards of knowing that it contributed to significant improvements to the community. Its rewards are seeing families picnicking in the park, children playing on the playground, and teens using the new ball fields. The members have the joy of creatively envisioning the future of the park system and of getting a notice of a grant application being awarded.

The National Park Service has been using volunteers for years. In 1970 the Volunteers In Parks (VIP) program was authorized so that the National Park Service could accept and utilize voluntary help and services from the public.

Each year more than 120,000 volunteers donate over 4 million hours of service in the U.S. national parks. Volunteers from every state and many countries around the world, from all walks of life and all age groups, come to help preserve and protect America's natural and cultural heritage for the enjoyment of this and future generations. Not-for-profit organizations are made up of volunteers and have missions to follow. If their mission is complementary to a part of the park board's mission, it's a natural fit for the development of a partnership. Not-for-profits may need little or no direction. If it is their responsibility to maintain a safe and orderly trail, they will organize the operation, schedule the volunteer activities, and find the funding to buy needed equipment.

There are just a few basic qualities that characterize a good volunteer event or activity:

- The more the merrier. Invite more hands than you need. People like to work with others. Make it a social event. Make it fun.

- Identify a job for everyone. Physical strength isn't everything. Some would rather organize, some have only the strength or interest to serve drinks. Some want to work hard and sweat, and others would rather record the event on film.
- Show them the money. Show the volunteers how valuable their time and service is to the community. Show them how their contribution is used as a match for a grant or for savings for the park board.
- Give them food. People don't come just for the food, but food is a great reward. Something as simple as cold lemonade tastes like a treat when you're hot and tired.
- Get them started young and keep them engaged. Teach kids the value of volunteerism, and they will carry that ethic for life. However, all ages can get involved. Some outstanding volunteers start when they retire from their professional lives.

Figure 10-1B

Volunteers can't do everything. It is important to be able to distinguish between the feasible volunteer activities and those that should be done by the professionals. But never underestimate the power of a cleverly delivered challenge. Many people are more challenged by the suggestion that they can't do something difficult than by the request to do something easy. Volunteers must be managed to the best advantage of the parks and themselves. Tom Sawyer used volunteers to paint the fence.

Clarity of vision is of utmost importance in producing a great park system and attracting money and supporters. The best design is only a useless dream without the money and organization to build it. The key to attracting funding through grants, sponsorships, partnerships, or technical assistance is to have big plans and dreams. Creative and innovative plans attract attention, are newsworthy, and engender the enthusiasm of the community and potential supporters. Everyone wants to be part of a winning idea; everyone wants to be part of the bright and beautiful.

Following is a case study of a small Midwestern town that has been awarded grants, built partnerships, nurtured sponsorships, and developed a huge volunteer network.

The story starts in Delphi, a small town located on the banks of the Wabash River and historically traversed by the Wabash and Erie Canal. An historical interest group adopted the canal as a preservation project and became known as the Canal Association, a not-for-profit corporation. Early in its existence the association had received a gift of land from an industry close to the canal. Thus began the Canal Park's early development, with a collection of canal-era cabins, outbuildings, and a vision sustained over the next two decades.

Some members of the association wanted to go farther and faster with a fund-raising campaign to dredge and restore the canal. Twelve years earlier, the association had a study done by a professional fundraiser who concluded that there just wasn't enough money in the small community to donate the kind of revenue they needed.

Almost simultaneously, a member of the association contacted a professor at a nearby university program in landscape architecture. After discussing the needs and goals of the organization, the two decided to conduct a student design project on the park (technical assistance from a university). In the course of the design project, students discovered that the community as a whole had a much larger opportunity for enhancing the park system on a different piece of property at the edge of town, near the river and canal. It just so happened that the site had been targeted for a new car crushing/metal recycling plant. The student project sparked some new ideas and generated excitement in the association. The community rallied to stop the industrial development at the entrance to town. The professor notified the association of a new grant program in the state Department of Natural Resources targeted at the resources of the Wabash River. A big idea had been revitalized.

The only eligible recipients for the grants were identified as bona fide park boards. A partnership formed between the park board and the Canal Association. This partnership was born out of necessity but has lasted for years and is alive and well at the time of this writing. The park board signed on as the project sponsor and took on the legal responsibility, the association committed to raising the local match for the grant, and a group of local organizations hired a landscape architect to prepare the grant application and preliminary site plans for the new park and trail.

The first of several grants was for the purchase of the new parkland, the development of the park, and the development of the first phase of a trail system. The trail would parallel the canal. The Canal Association had received a short section of the canal as a gift from a local donor. The land ownership was transferred from the association to the park board, which constituted part of the in-kind donation of land. The association used this land and the pledge of donated labor as an in-kind grant match. Subsequent grants with associated in-kind matches of land and labor helped create a large part of the total trail system and various facilities along the canal.

The association also used a local community foundation grant as a match for a state grant program. Match requirements are usually clearly spelled out in the grant application. If in doubt, ask the grant administrator. However, grants can be used to leverage other grants. In the end, there is almost always a requirement for some input of the recipient's money. In the case of the association's plan to build an interpretive center, a local grant was used to leverage a state grant, and both of those grants were used to match a federal grant.

Various legal services were donated by a retired lawyer. This lawyer, through his own curiosity, discovered several rights of ownership that had gone unnoticed. His meticulous research and note taking saved or earned the association thousands of dollars in legal fees and land ownership. (Note that a lawyer volunteering to pick up sticks or operate a shovel is an in-kind labor at minimum-wage rates, whereas a lawyer volunteering to do legal work for a park board or organization is contributing at a considerably higher rate. This could be called a sponsorship. But don't turn away a lawyer who wants to help pick up sticks.)

The Canal Association sought and received several kinds of technical assistance from the National Park Service Rivers, Trails, and Conservation Assistance (RTCA) program (technical assistance from a government agency). Rory Robinson, the RTCA staff person assigned to Indiana, assisted Delphi in conducting a community visioning process that helped them identify and prioritize their goals and define specific actions for development. The community also was provided valuable information about various funding programs and even offered a challenge grant to help out on a small project. RTCA also provided direct assistance on the design of an interpretive signage system and help the association select an interpretive specialist.

The Canal Association has established a volunteer network that is the envy of many cities much greater in population. The volunteer team starts with Dan McCain, a retired soil conservationist, is complemented by a core of local retired farmers and professionals (even an airline pilot), and is expanded by a cadre of various individuals with all kinds of skills to offer. The volunteers are involved in all the planning; they help set priorities and schedule workdays. And the workday schedule is one that everyone can remember—the third Saturday of every month. In some months with inclement weather, they just have a fun day, such as an organized hike. The big picture is always evident. Twice a year they have a big workday with lunch provided. (Give 'em food.) The food is donated and served by a local service sorority. Everybody gets in on the fun, and everyone has a meaningful role to play. The major workdays typically begin with a ceremony of dedication or memorial and usually include the Girl Scouts and Boy Scouts. Some projects are carried out as Eagle Scout projects. Everyone signs in so that there is always a record of the event, the volunteers, and the number of hours that they contribute. They know that their contributions are integral parts of the vision—the big picture. And the annual volunteer appreciation day is fun for everyone.

This is not the full story; it just illustrates some of the ways in which the community of Delphi and the Canal Association have used just about all the available techniques for assembling the funds and resources to build parks, trails, and more. The best part is that everyone who has helped in any way has assumed a sense of ownership of the various facilities. So can you purchase a security system to protect the parks and trails? Yes, but you get a pretty good security system for free when everyone in the community wants to protect their investment. They worked hard to build it; they aren't going to let anyone damage it.

They did the planning, they did the preparation, they got the funding and the resources, and they did the work—and they had more than a park when they were done. But they're still not finished. There will always be more to come as people change and ideas continue to be generated. Other towns or clients may only need to do one of these things, but *everybody* can do *any* of them.

So go on out and have yourself a park.

Appendices

The following information illustrates the kinds of empirical data and tools available to recreation area designers. Approach it with the precautions expressed in chapters 2 and 4 in mind. More extensive tables related to these and other aspects of design are found in many of the books listed in the bibliography.

Appendix 1

Selected Park and Activity Size and Facility Standards

Tot Lot (2400 to 5000 square feet)
Usually includes: chair swings, sandbox, regular swings, slide, climbing apparatus, wading or spray pool, playhouse, turf area, paved area for wheeled toys, benches.

Neighborhood Playground (2.5 to 10 acres)
Usually includes: play apparatus, turf area, paved court, playfield, storytelling ring, shelter, wading or spray pool, table-game area, picnic center.

Neighborhood Park (2 to 5 acres)
Usually includes: open lawn, trees, shrubbery, walks, benches, focal point such as ornamental pool or fountain, sandbox, play apparatus, table-game area.

Community Playfield (15 to 25 acres)
Usually includes: separate sport fields for men and women; courts for tennis, boccie, horseshoes, shuffleboard, and other court games; lawn areas for croquet, archery, and other lawn sports; outdoor swimming pool, band shell, family picnic area, children's playground, running track, day camp center, parking area.

City Park (100 to 200 acres)
Usually includes: facilities for boating, swimming, picnicking, hiking, field sports, day camp, zoo, arboretum, nature museum.

County Park (200 acres or more)
Usually includes: preschool play area (3 acres), elementary play area (4 acres), sport fields (15 acres), paved courts (3 acres), mul-

tiple-use court (1 acre), family picnic area (30 acres), open area for special events (10 acres), amphitheater (7 acres), natural area (40 acres), parking for 2000 cars (15 acres), day and weekend camping area (25 acres), clubhouse and recreation center (5 acres), maintenance yard (5 acres), landscaped area (15 acres), service roads (20 acres).

Natural Environment Area (5-acre minimum)
Usually includes: picnic area (8 acres), tent camping area (6 acres), trailer camping area (6 acres), hiking trails, boat access, marina, sightseeing facilities, parking areas.

Nine-Hole Golf Course (75 acres—double for eighteen holes)
Usually includes: fairways, roughs, greens, and tees (43 acres), clubhouse (.25 acre), parking area and service roads (1.75 acres), natural area (20 acres), landscaped area (10 acres).

Lake or Ocean Swimming
For every 25 linear feet of shoreline: 5000 square feet for sunbathing, 2500 square feet for buffer and picnicking, 1000 square feet of water area for swimming.

Nature Trail (1 to 2 miles long each)
Fifty people per mile of trail per day.

Rural Hiking Trail
Forty people per mile of trail per day.

Urban Hiking Trail
Ninety people per mile of trail per day.

Family Picnic Area within Community
Sixteen units per acre, each unit consisting of table and cooking facilities. One off-street car space per unit.

Family Picnic Area outside Community
Ninety to one hundred twenty units per area; ten to fifteen units per acre, each unit consisting of one table-bench combination with one charcoal-burning stove per two tables; one comfort station per each thirty units serving a 500-foot radius; drinking fountains no further than 150 feet from picnic units; garbage cans in racks near circulation road no further than 150 feet from picnic units, but not near drinking fountain; bulletin board near comfort station.

Campground
Ninety to one hundred twenty units per area; four to seven units per acre consisting of one tent area, one table-bench combination, one camp stove, one parking space; one comfort station per each thirty campsites serving a 300-foot radius; drinking fountains, garbage cans, and bulletin board as for picnicking, above.

Appendix 2

Selected Game Area Size Standards

Name	Dimensions of Game areas	Use Dimensions (linear feet)	Space Required (square feet)
Archery	90′–300′ in length Targets 15′ apart	50 × 175 (min.) 50 × 400 (max.)	8,750 20,000
Badminton	17′ × 44′ (singles) 20′ × 44′ (doubles)	25 × 60 30 × 60	1,500 1,800
Baseball	90′ diamond	350 × 350 (average with hooded backstop) 400 × 400 (without backstop)	122,500 160,000
Basketball	42′ × 74′ (min.) 50′ × 94′ (max.)	60 × 100 (average)	6,000
Boccie	8′ × 62′	20 × 80	1,600
Clock Golf	20′–30′ diameter	40 × 40	1,600
Croquet	41′ × 85′	50 × 95	4,750
Curling	Tees 114′ apart	25 × 160	4,000
Field hockey	150′ × 270′ (min.) 180′ × 300′ (max.)	210 × 330 (average)	69,300
Football	160′ × 360′	200 × 420	84,000
Handball	20′ × 34′	30 × 45	1,350
Horseshoes	Stakes 40′ apart	12 × 52 (min.)	624
Lacrosse	180′ × 330′ (min.) 210′ × 330′ (max.)	225 × 360 (average)	81,000
Lawn bowling	141′ × 110′ (1 alley)	130 × 130	16,900
Shuffleboard	6′ × 52′	10 × 60	600
Soccer	165′ × 300′ (min.) 225′ × 360′ (max.)	225 × 360 (average)	81,000
Softball	55′ diamond	275 × 275 (min.)	75,625
Tennis	27′ × 78′ (singles) 36′ × 78′ (doubles)	50 × 120 60 × 120	6,000 7,200
Volleyball	30′ × 60′	45 × 80	3,600

For more information on amateur sports field and court layout details, as well as rules of play and organization, contact the following references:

College-level sports, general
The National Collegiate Athletic Association (NCAA)
700 W. Washington Street
P.O. Box 6222
Indianapolis, IN 46206-6222

High school-level sports, general

National Federation of High School Athletic Associations (NFHS)
P.O. Box 690
Indianapolis, IN 46206

Amateur sports, general

Amateur Athletic Union-U.S. (AAU)
AAU National Headquarters
1910 Hotel Plaza Blvd.
P.O. Box 22409
Lake Buena Vista, FL 32830

Softball

Amateur Softball Association
2801 NE 50th St.
Oklahoma City, OK 73111

Pony League Baseball/Softball

Pony Baseball and Softball International Headquarters
P.O. Box 225
Washington, PA 15301-0225

Little League Baseball

Website—www.littleleague.org

Eastern Region Headquarters
P.O. Box 2926
Bristol, CT 06011

Central Region Headquarters
9802 E. Little League Drive
Indianapolis, IN 46235

Southern Region Headquarters
P.O. Box 13366
St. Petersburg, FL 33733

Western Region Headquarters
6707 Little League Drive
San Bernardino, CA 92407

Southwestern Headquarters
P.O. Box 20127
Waco, TX 76702

Canada Headquarters
235 Dale Avenue
Ottawa, Ontario
Canada K1G 0H6

Appendix 3

Field and Court Games

The drawings shown here are for planning use as general templates, not for detail of layout. Soccer is shown at the largest (championship) size. Rugby is shown with the widest of multiple references but sometimes is given a length up to 30 yards longer. Verify details from the resource agencies given in Appendix 2.

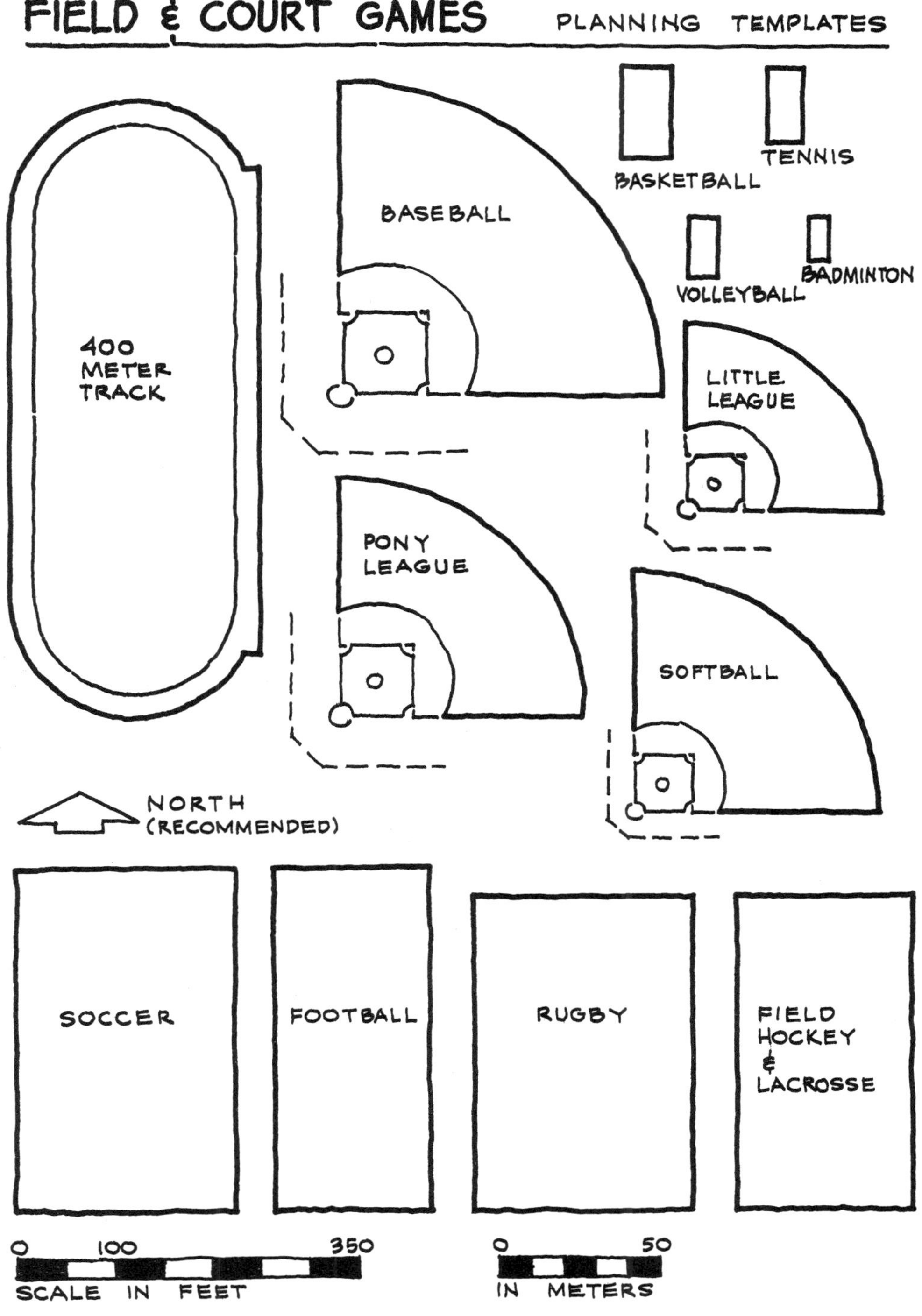

Appendix 4

Responses of Selected Trees to Recreational Use

Abstracted from the Southeastern Forest Experiment Station Research Note 171, Asheville, North Carolina, February 1962, the following represents the results of research conducted on forty-two camping and picnicking sites on the Cherokee, Nantahala, and Pisgah National Forests by the United States Department of Agriculture, Forest Service, Thomas H. Ripley, Project Director.

Data indicated little or no difference in damage or in insect and disease problems related to size or dominance of the vegetation. Differences between species, however, were consistently large. The following conifers and hardwoods are listed in order of decreasing ability to withstand the impacts of recreation use, as gauged by disease infection, insect infestation, and decline.

Conifers

1. Shortleaf pine
2. Hemlock
3. White pine
4. Pitch pine
5. Virginia pine

Hardwoods

1. Hickories
2. Persimmon
3. Sycamore
4. White ash
5. Beech
6. Sassafras
7. Buckeye
8. Yellow poplar
9. Dogwood
10. Blackgum
11. Yellow birch
12. Red maple
13. American holly
14. Sourwood
15. Black birch
16. White oaks
17. Black walnut
18. Red oak
19. Black locust
20. Magnolia
21. Black cherry
22. Blue beech

Conifers were clearly more susceptible to disease and insect attack than were hardwoods—with the possible exception of shortleaf pine and hemlock. Plant material guides for selection of appropriate vegetation for park and recreational use in other regions of the United States can be found in the Bibliography.

**Regarding any plant selections:* Virtually all plants are susceptible to diseases and pest damage. Problems with any of even the most popular plants will occur unpredictably. Before making selections and major investments, all plant choices should be reviewed by consulting with local experts and extensive web-based searches for relevant, current information.

Appendix 5

Suitability of Soils for Recreational Use

The following pages are examples of soil limitations that affect their suitability for recreation. The ratings are based on restrictive soil features such as wetness, slope, and texture of the surface layer. Susceptibility to flooding is also considered. Tables such as these are included in all current soil surveys from county offices of the Natural Resources Conservation Service. The examples included here are from Tippecanoe County, Indiana.

This soil survey is a publication of the United States Department of Agriculture, Natural Resources Conservation Service prepared in cooperation with Purdue University Agriculture Experiment Station and the Indiana Department of Natural Resources, State Soil Conservation Board, and Division of Soil Conservation. It was issued in 1998. All states prepare soil surveys in similar cooperative arrangements. They are available at local county NRCS offices. Soil surveys should be used in conjunction with, not instead of, careful on-site inspection of soils and other site resources. Soil surveys often include additional useful information such as soil suitability to construction of buildings and septic systems, road building, wildlife habitat, and water management.

Additional soils information is available from the following organizations:

Natural Resource Conservation Service
The American Society of Agronomy
The Soil Science Society of America

Additionally, state agricultural universities and departments of natural resources may have additional resources on their websites.

Tippecanoe County, Indiana
Table REC-1.--Recreation

Print date: 05/07/2002

(The information in this table indicates the dominant soil condition but does not eliminate the need for onsite investigation. The numbers in the value columns range from 0.01 to 1.00. The larger the value, the greater the limitation. See text for further explanation of ratings in this table.)

Map symbol and soil name	Pct. of map unit	Camp areas		Picnic areas		Playgrounds	
		Rating class and limiting features	Value	Rating class and limiting features	Value	Rating class and limiting features	Value
Am:							
Allison-------------	90	Very limited		Not limited		Not limited	
		Flooding	1.00				
Ap:							
Allison-------------	90	Very limited		Somewhat limited		Very limited	
		Flooding	1.00	Flooding	0.40	Flooding	1.00
AtB2:							
Alvin---------------	60	Not limited		Not limited		Somewhat limited	
						Slope	0.50
Spinks--------------	30	Somewhat limited		Somewhat limited		Somewhat limited	
		Too sandy	0.88	Too sandy	0.88	Too sandy	0.88
						Slope	0.50
Ba:							
Battleground--------	100	Very limited		Not limited		Not limited	
		Flooding	1.00				
Bb:							
Battleground--------	90	Very limited		Somewhat limited		Very limited	
		Flooding	1.00	Flooding	0.40	Flooding	1.00
BgA:							
Beecher-------------	90	Very limited		Somewhat limited		Very limited	
		Depth to saturated zone	1.00	Depth to saturated zone	0.94	Depth to saturated zone	1.00
		Restricted permeability	0.21	Restricted permeability	0.21	Restricted permeability	0.21
BkF:							
Berks---------------	100	Very limited		Very limited		Very limited	
		Slope	1.00	Slope	1.00	Slope	1.00
		Gravel content	0.08	Gravel content	0.08	Gravel content	1.00
						Content of large stones	0.68
						Depth to bedrock	0.42
BlA:							
Billett-------------	100	Not limited		Not limited		Not limited	
BlB2:							
Billett-------------	100	Not limited		Not limited		Somewhat limited	
						Slope	0.50
BmA:							
Billett-------------	100	Somewhat limited		Somewhat limited		Somewhat limited	
		Depth to saturated zone	0.07	Depth to saturated zone	0.03	Depth to saturated zone	0.07
BnA:							
Billett-------------	100	Not limited		Not limited		Not limited	
BnB2:							
Billett-------------	100	Not limited		Not limited		Somewhat limited	
						Slope	0.50
BoA:							
Bowes---------------	90	Not limited		Not limited		Not limited	
BpA:							
Bowes Variant-------	90	Somewhat limited		Somewhat limited		Somewhat limited	
		Depth to saturated zone	0.07	Depth to saturated zone	0.03	Depth to saturated zone	0.07
CaA:							
Camden--------------	100	Not limited		Not limited		Not limited	
CfB:							
Carmi---------------	100	Not limited		Not limited		Somewhat limited	
						Slope	0.50
						Gravel content	0.06

Tippecanoe County, Indiana
Table REC-2.--Recreation

Print date: 05/07/2002

(The information in this table indicates the dominant soil condition but does not eliminate the need for onsite investigation. The numbers in the value columns range from 0.01 to 1.00. The larger the value, the greater the limitation. See text for further explanation of ratings in this table.)

Map symbol and soil name	Pct. of map unit	Paths and trails		Off-road motorcycle trails		Golf fairways	
		Rating class and limiting features	Value	Rating class and limiting features	Value	Rating class and limiting features	Value
Am:							
Allison	90	Not limited		Not limited		Not limited	
Ap:							
Allison	90	Somewhat limited		Somewhat limited		Very limited	
		Flooding	0.40	Flooding	0.40	Flooding	1.00
AtB2:							
Alvin	60	Not limited		Not limited		Not limited	
Spinks	30	Somewhat limited		Somewhat limited		Somewhat limited	
		Too sandy	0.88	Too sandy	0.88	Droughty	0.62
Ba:							
Battleground	100	Not limited		Not limited		Not limited	
Bb:							
Battleground	90	Somewhat limited		Somewhat limited		Very limited	
		Flooding	0.40	Flooding	0.40	Flooding	1.00
BgA:							
Beecher	90	Somewhat limited		Somewhat limited		Somewhat limited	
		Depth to saturated zone	0.86	Depth to saturated zone	0.86	Depth to saturated zone	0.94
BkF:							
Berks	100	Very limited		Very limited		Very limited	
		Slope	1.00	Slope	1.00	Slope	1.00
						Droughty	0.95
						Content of large stones	0.68
						Depth to bedrock	0.42
						Gravel content	0.08
BlA:							
Billett	100	Not limited		Not limited		Not limited	
BlB2:							
Billett	100	Not limited		Not limited		Not limited	
BmA:							
Billett	100	Not limited		Not limited		Somewhat limited	
						Depth to saturated zone	0.03
BnA:							
Billett	100	Not limited		Not limited		Not limited	
BnB2:							
Billett	100	Not limited		Not limited		Not limited	
BoA:							
Bowes	90	Not limited		Not limited		Not limited	
BpA:							
Bowes Variant	90	Not limited		Not limited		Somewhat limited	
						Depth to saturated zone	0.03
CaA:							
Camden	100	Not limited		Not limited		Not limited	
CfB:							
Carmi	100	Not limited		Not limited		Not limited	
CgA:							
Carmi	100	Not limited		Not limited		Not limited	
Ck:							
Ceresco	90	Somewhat limited		Somewhat limited		Somewhat limited	
		Depth to saturated zone	0.86	Depth to saturated zone	0.86	Depth to saturated zone	0.94

Appendix 6

Agencies and Organizations Offering Technical Assistance

Chapter 9 refers to several sources of technical assistance from government agencies, corporations, national or state not-for-profits, universities, and knowledgeable individuals. There are university programs in landscape architecture, landscape management, recreation studies, planning, and architecture that offer technical assistance in every aspect of park planning, design, management, and funding. There are too many to list, but they are not difficult to find. Corporations, state not-for-profits, and individuals are far too numerous to include them all here, but the following lists will represent a starting point in a search for sources of assistance.

National Not-for-Profit Organizations

American Society of Landscape Architects*
American Forests
American Rivers
American Trails
The Conservation Fund
National Parks and Conservation Association
National Recreation and Parks Association**
National Trust for Historic Preservation
The Nature Conservancy
Rails to Trails Conservancy
Surface Transportation Policy Project
The Trust for Public Land
The Wilderness Society

Federal Agencies

Bureau of Land Management
Center of Excellence for Sustainable Development
Federal Highway Administration
National Park Service
Rivers, Trails and Conservation Assistance
Natural Resource Conservation Service
US Army Corps of Engineers
USDA–Forest Service
U.S. Department of Energy
U.S. Fish and Wildlife Service

*Universities and colleges with accredited programs in landscape architecture are also listed by ASLA.
**Also has a full list of accredited programs in recreation and park administration.

Bibliography

The numbers in parentheses that follow each citation refer to the chapters in this book which the reference most directly augments. An asterisk (*) indicates that the reference contains information related to the natural resource aspects of park system planning.

AASHTO Task Force on Geometric Design, *Guide for the Development of Bicycle Facilities*. American Association of State Highway and Transportation Officials, Washington, DC, 1999. (6)

Aaron, David, with Bonnie P. Winawer, *Child's Play: A Creative Approach to Playscapes for Today's Children*. Harper & Row, New York, 1965. (2, 3)

Allen, Marjorie, *Design for Play*. E. T. Hawn and Company, Ltd., London, 1962. (2, 3)

———, *Planning for Play*. MIT Press, Cambridge, MA, 1968. (2, 3)

Agee, James K., and Darryll R. Johnson, *Ecosystem Management for Parks and Wilderness (Contribution)*. University of Washington Press, Seattle, 1989. (4, 7, *)

Alef, Daniel, *Frederick Law Olmsted: Mogul of Parks and Landscaping (Titans of Fortune)*. Meta4 Publishing, Santa Barbara, CA, 2009. (1)

Appleyard, Donald, Kevin Lynch, and John R. Myer, *The View from the Road*. MIT Press, Cambridge, MA, 1964. (2)

Ardrey, Robert, *The Territorial Imperative*. Atheneum, New York, 1966. (2)

Arendt, Randall, *Growing Greener: Putting Conservation into Local Plans and Ordinances*. Natural Lands Trust/Island Press, 1999. (9)

Barlow Rogers, Elizabeth, *Landscape Design: A Cultural and Architectural History*. New York: Harry N. Abrams, New York, 2001. (1, 2, 3)

Bauer, David G., *The "How To" Grants Manual: Successful Grant-seeking Techniques for Obtaining Public and Private Grants*. American Council on Education, Macmillan, New York, 1988. (9)

Blake, Peter, *God's Own Junkyard: The Planned Deterioration of America's Landscape*. Holt, Rinehart, and Winston, New York, 1964. (1)

Brash, Alexander, Jamie Hand, and Kate Orff. *Gateway: Visions for an Urban National Park*. Princeton Architectural, New York, 2011. (2, 3, 4)

Butler, George D., *Introduction to Community Recreation*. McGraw-Hill, New York, 1976. (2, 4)

Calkins, Meg. *The Sustainable Sites Handbook: A Complete Guide to the Principles, Strategies, and Best Practices for Sustainable Landscapes*. John Wiley & Sons Canada, Limited, Toronto, Ontario, 2011. (2, 4, *)

Campbell, Craig S., and Michael Ogden. *Constructed Wetlands in the Sustainable Landscape*. Wiley, New York, 1999. (3, 4, *)

Carr, Shaun Eyring, and Richard Guy Wilson, *Public Nature: Scenery, History, and Park Design*. University of Virginia Press, Charlottesville, 2013. (1, 2, 3)

Colley, David P., and Elizabeth Keegin Colley, *Prospect Park: Olmsted and Vaux's Brooklyn Masterpiece*. Princeton Architectural, New York, 2013. (1)

Cornish, Geoffrey S., and Robert Muir Graves, *Golf Course Design*. John Wiley & Sons, New York, 1998. (2, 3, 4)

Cremin, Dennis H., *Grant Park: The Evolution of Chicago's Front Yard*. Board of Trustees, Southern Illinois University, Carbondale, 2013. (1)

Crompton, John L. *The Proximate Principle: The Impact of Parks, Open Space and Water Features on Residential Property Values and the Property Tax Base*. Ashburn, VA: National Recreation and Park Association, 2004. (3, 4, 7, 9)

Czerniak, Julia, and George Hargreaves. *Large Parks*. New York: Princeton Architectural, 2007. (3, 4)

David, and Robert Hammond. *High Line: The Inside Story of New York City's Park in the Sky*. FSG Originals, New York, 2011. (1, 9)

Diekelmann, John, and Robert Schuster, *Natural Landscaping: Designing with Native Plant Communities*. McGraw-Hill, New York, 1982. (4, 6, *)

Dines, Nicholas T., and Kyle D. Brown, *Landscape Architect's Portable Handbook*. McGraw-Hill, New York, 2001. (2)

Dirr, Michael A., *Dirr's Encyclopedia of Trees and Shrubs*. Timber Press, Portland, OR, 2011. (3, 4, *)

———, *Dirr's Trees and Shrubs for Warm Climates: An Illustrated Encyclopedia*. Timber Press, Portland, OR, 2002. (3, 4, *)

Doell, Charles E., and Louis F. Twardzik, *Elements of Park and Recreation Administration*. Burgess, Minneapolis, MN, 1979. (1, 2, 4, 6)

Douglass, Robert W., *Forest Recreation*, 5th Ed. Waveland Press, Long Grove, IL, 2000. (4)

Dubos, René, *So Human an Animal: How We Are Shaped by Surroundings and Events*. Charles Scribner's Sons, New York, 1969. (2, 3)

Dulles, Rhea D., *A History of Recreation*. Meredith, Des Moines, IA, 1965. (1)

Eckbo, Garrett, *The Landscape We See*. McGraw-Hill, New York, 1969. (1, 2, 3, 4, 5, 6,*)

Eckbo, Garrett, et al., *People in the Landscape*. Prentice-Hall, Upper Saddle River, NJ, 1998. (1, 2, 3, 4, 5, 6,*)

Ellis, Michael J., *Why People Play*. Prentice-Hall, Englewood Cliffs, NJ, 1973. (2)

Ellis, Michael J., and G. J. L. Scholtz, *Activity and Play of Children*. Prentice-Hall, Englewood Cliffs, NJ, 1978. (2)

Fabos, Julius G., Gordon T. Milde, and Michael Weinmayer, *Frederick Law Olmsted, Sr.: Founder of Landscape Architecture in America*. University of Massachusetts Press, Amherst, 1968. (1)

Ferguson, Bruce K. *Porous Pavements*. CRC, Boca Raton, FL, 2005. (4, 6, *)

Fine, Albert, *Landscape into Cityscape: Frederick Law Olmsted's Plans for a Greater New York*. Cornell University Press, Ithaca, NY, 1968. (1)

Flink, Charles A., Robert M. Searns, and Kristine V. Olka, *Trails for the Twenty-First Century*. Rails-to-Trails Conservancy, Island, Washington, DC, 2001. (6)

Flint, Harrison L., *Landscape Plants for Eastern North America*. WileyInterscience, New York, 1983. (4)

Forsyth, Ann, and Laura Musacchio, *Designing Small Parks: A Manual for Addressing Social and Ecological Concerns*. John Wiley & Sons, Hoboken, NJ, 2005. (2, 3, 4, *)

Garvin, Alexander, *Public Parks: The Key to Livable Communities* (Norton/Library of Congress Visual Sourcebooks in Architecture, Design, and Engineering). W. W. Norton, New York, 2010. (1, 2)

Hall, Edward T., *The Hidden Dimension*. Doubleday & Company, Garden City, NY, 1966. (3)

———, *The Silent Language*. Greenwood, Westport, CT, 1980. (2)

Hanna, Karen C., *GIS for Landscape Architects*. ESRI, Redlands, CA, 1999. (1)

Harper, Jack. *Planning for Recreation and Parks Facilities: Predesign Process, Principles, and Strategies*. Venture, State College, PA, 2009. (2, 3, 4, 6, 7)

Harris, Charles W., and Nicholas T. Dines, *Time-Saver Standards for Landscape Architects*, 2nd Ed. McGraw-Hill, New York, 1998. (4, 5, 7,*)
Harvard University Landscape Architecture Research Office, *Three Approaches to Environmental Analysis*. The Conservation Foundation, Washington, DC, 1967. (6, *)
Hellmund, Paul Cawood., and Daniel S. Smith. *Designing Greenways: Sustainable Landscapes for Nature and People*. Island, Washington, DC, 2006. (3, 4, 6, *)
Holden, Robert, and Jamie Liversedge, *Landscape Architecture: An Introduction*. Laurence King, London, 2014. (2, 3, 4)
Hultsman, John, Richard L. Cottrell, and Wendy Z. Hultsman, *Planning Parks for People*. Venture, Andover, ME, 1998. (2, 3, 6)
Kaplan, R., and S. Kaplan, *The Experience of Nature: A Psychological Perspective*. Cambridge University Press, New York, 1989. (2)
Kaplan, R., S. Kaplan, and R. L. Ryan, *With People in Mind: Design and Management of Everyday Nature*. Island, Washington, DC, 1998. (2, *)
Kowsky, Francis R., *The Best Planned City in the World: Olmsted, Vaux, and the Buffalo Park System (Designing the American Park)*. Library of American Landscape History, Amherst, MA, 2013. (1, 2)
Krauel, Jacobo, *New City Parks*. Carles Broto, Barcelona, Spain, 2008. (1, 2, 9)
Ledermann, Alfred, and Alfred Trachsel, *Creative Playgrounds and Recreation Centers*. Frederick A. Praeger, New York, 1968. (2, 3)
Larabee, Jonathan M., *How Greenways Work: a Handbook on Ecology*. National Park Service and Atlantic Center for the Environment, Ipswich, MA, 1992. (6)
Lewis, Philip H., Jr., *Tomorrow by Design: A Regional Design Process for Sustainability*. John Wiley & Sons, New York, 1968. (*)
Little, Charles E., *Greenways for America*. The Johns Hopkins Press, Baltimore, 1990. (6)
Low, Setha Dana Taplin and Suzanne Scheld, *Rethinking Urban Parks: Public Space and Cultural Diversity*. University of Texas Press, Austin, 2005. (1, 2)
Lynch, Kevin, and Gary Hack, *Site Planning*, 3rd Ed. MIT Press, Cambridge, MA, 1984. (2, 3, 4, 5, 6)
Marcus, Clare Cooper, and Carolyn Francis (Eds.); *People Places: Design Guidelines for Urban Open Space,* 2nd Ed. Wiley, New York, 1997. (2, 3, 4)
Marshall, Lane L., *Action by Design*. American Society of Landscape Architects, Washington, DC, 1983. (1)
McHarg, Ian L., *Design with Nature*. The Natural History Press, Garden City, NY, 1969. (1, 2, 3, 4, 6, *)
Melind, Tom, *Park Doctor: How to Revitalize Your Park*. Author House, Bloomington, IN, 2006. (2, 3, 4, 6)
Molnar, Donald J., "Physical Design Criteria for the Landscape Design of Preschool Play Areas." Master's Thesis, University of Illinois, Urbana, 1964. (2, 3, 7)
Moore, Robin C., Susan M. Goltsman, and Daniel S. Iacofano. *Play for All Guidelines: Planning, Design, and Management of Outdoor Play Settings for All Children*. MIG Communications, Berkeley, CA, 1992. (2, 3, 4, 6, 7)
Morrow, Baker H., *Best Plants for New Mexico Gardens and Landscapes: Keyed to Cities and Regions in New Mexico and Adjacent Areas*. University of New Mexico Press, Albuquerque, 1995. (4)
Olgay, Victor, *Design with Climate*. Princeton University Press, Princeton, NJ, 1963. (4)
Pena, William M., with William W. Caudill and John W. Focke, *Problem Seeking*. Cahners Books International, Boston, 1977. (6)
Phillips, Leonard E., *Parks: Design and Management*. McGraw-Hill, New York, 1996. (2,*)
PLAE, Inc. et al., *Universal Access to Outdoor Recreation*. PLAE, Inc. Berkeley, CA, 1993. (4)
Platt, Rutherford H. (Ed.), *The Humane Metropolis: People and Nature in the Twenty-first Century City*. The Lincoln Institute of Land Policy Cambridge, ME, 2006. (1, 2)

Ripley, T. H., *Tree and Shrub Response to Recreation Use*. U.S. Forest Service, Southeastern Forest Experiment Station, Research Note 171, Asheville, NC, 1962. (4)
Rubenstein, Harvey M., *A Guide to Site and Environmental Planning,* 3rd Ed. John Wiley & Sons, New York, 1987. (2, 3, 4, 5, 6)
Rutledge, Albert J., *A Visual Approach to Park Design* (Garland series in design). Garland, New York, 1985. (2, 3, 4, 9)
Ryan, R. L., R. Kaplan, and R. E. Grese, "Predicting Volunteer Commitment in Stewardship Programs." *Journal of Environmental Planning and Management*, Vol. 44, No. 5, 629–648, 2001. (9)
Simonds, John O., *Earthscape: A Manual of Environmental Planning*. John Wiley & Sons, New York, 1997. (1)
———, *Landscape Architecture: A Manual of Site Planning and Design*, 3rd Ed. McGraw-Hill, New York, 1997. (1, 2, 3, 4, 5, 6)
Sommer, Robert, *Personal Space: The Behavioral Basis of Design*. Prentice-Hall, Englewood Cliffs, NJ, 1969. (2)
Starke, Barry, and John Ormsbee Simonds, *Landscape Architecture: A Manual of Environmental Planning and Design*, 5th Ed. McGraw-Hill Education, Columbus, OH, 2013. (1, 2, 3)
Udall, Stewart L., *The Quiet Crisis*. Holt, Rinehart, and Winston, New York, 1963. (1,*)
Urban, James. *Up by Roots: Healthy Soils and Trees in the Built Environment*. International Society of Arboriculture, Champaign, IL, 2008. (3, 4, 6, *)
Van Meter, Jerry R., *Master Plans for Park Sites*. University of Illinois Cooperative Extension Service, Urbana, 1969. (2, 4, 6)
Weddle, A. E. (Ed.), *Techniques of Landscape Architecture*. William Heinemann, Ltd., London, 1967. (2, 3, 4, 5, 6)
Wheeler, Stephen M., *Planning for Sustainability: Creating Livable, Equitable and Ecological Communities*. Routledge, London/New York, 2013. (2,*)
Whyte, William H., *The Last Landscape*. Doubleday & Company, Garden City, NY, 1968. (1,*)
———, *The Social Life of Small Urban Spaces*. The Conservation Foundation, Washington, DC, 1980. (2)
Wyman, Donald, *Shrubs and Vines for American Gardens*, Rev. Ed. Hungry Minds, New York, 1996. (4)
———, *Trees for American Gardens*, 3rd Ed. Macmillan, New York, 1990. (4)

Online Sources

The American Camping Association website (http://www.acacamps.org) provides information primarily about organized group camps and camping. (4)
The American Planning Association website (http://www.planning.org) has information on planning conferences and publications related to parks, greenways, open space, and related planning issues. (2, 4, 8)
The American Society of Landscape Architects website (http://www.asla.org) has information on membership, selecting a landscape architect, and publications. ASLA has a professional interest group on recreation. (all chapters, *)
The National Recreation and Parks Association website (http://www.nrpa.org) has information on membership, conferences, seminars, and publications. (2, 4, 7, 9, 10)
The Urban Land Institute website (http://www.uli.org) provides access to research, conferences, and publications on parks, resorts, and other land-use issues. (2, 4, 8, 9)

Index